MW00908877

The Doctrine of Chances

You are holding a reproduction of an original work that is in the public domain in the United States of America, and possibly other countries.You may freely copy and distribute this work as no entity (individual or corporate) has a copyright on the body of the work.This book may contain prior copyright references, and library stamps (as most of these works were scanned from library copies).These have been scanned and retained as part of the historical artifact.

This book may have occasional imperfections such as missing or blurred pages, poor pictures, errant marks, etc. that were either part of the original artifact, or were introduced by the scanning process. We believe this work is culturally important, and despite the imperfections, have elected to bring it back into print as part of our continuing commitment to the preservation of printed works worldwide. We appreciate your understanding of the imperfections in the preservation process, and hope you enjoy this valuable book.

THE

DOCTRINE

CHANCES:

OR,

A Method of Calculating the Probability of Events in Play.

By *A. De Moivre.* F. R. S.

LONDON:

Printed by *W. Pearson*, for the Author. MDCCXVIII.

QA273
M65

Gift of J. Wargentin
to Math. Dept.

T O

Sir Isaac Newton, *Kt. President of the* Royal Society.

SIR

THE greatest help I have received in writing upon this Subject having been from your Incomparable Works, especially your Method of Series; I think it my Duty publickly to acknowledge, that the Improvements I have made in the matter here treated of, are principally derived from your self. The great benefit which has accrued to me in this respect, requires my share in the general Tribute of Thanks due to you from the Learned World: But one advantage, which is more particularly my own, is the Honour I have frequently had of being admitted to your private Conversation, wherein the doubts I have had upon any Subject relating to *Mathematics*, have been resolved by you with the greatest Humanity and Condescention. Those

Marks

Marks of your Favour are the more valuable to me, becaufe I had no other pretence to them, but the earneft defire of underftanding your fublime and univerfally ufeful Speculations. I fhould think my felf very happy, if, having given my Readers a Method of calculating the Effects of Chance, as they are the refult of Play, and thereby fix'd certain Rules, for eftimating how far fome fort of Events may rather be owing to Defign than Chance, I could by this fmall Effay excite in others a defire of profecuting thefe Studies, and of learning from your Philofophy how to collect, by a juft Calculation, the Evidences of exquifite Wifdom and Defign, which appear in the *Phænomena* of Nature throughout the Univerfe. I am, with the utmoft Refpect,

   Sir,

     Your moft Humble,

      and Obedient Servant,

        A. De Moivre.

PREFACE.

'TIS *now about Seven Years, since I gave a Specimen in the* Philosophical Transactions, *of what I now more largely treat of in this Book. The occasion of my then undertaking this Subject was chiefly owing to the Desire and Encouragement of the Honourable* Mr. Francis Robartes, *who, upon occasion of a French Tract, called,* L'Analyse des jeux de Hazard, *which had lately been Published, was pleased to propose to me some Problems of much greater difficulty than any he had found in that Book; which having solved to his Satisfaction, he engaged me to Methodise those Problems, and to lay down the Rules which had led me to their Solution. After I had proceeded thus far, it was enjoined me by the Royal Society, to communicate to them what I had discovered on this Subject, and thereupon it was ordered to be published in the Transactions, not as a matter relating only to Play, but as containing some general Speculations not unworthy to be considered by the Lovers of Truth.*

I had not at that time read any thing concerning this Subject, but Mr. Huygens's *Book, de Ratiociniis in Ludo Aleæ, and a little English Piece (which was properly a translation of it) done by a very ingenious Gentleman, who, tho' capable of carrying the matter a great deal farther, was contented to follow his Original; adding only to it the computation of the Advantage of the Setter in the Play called* Hazard, *and some few things more. As for the French Book, I had run it over but cursorily, by reason I had observed that the Author chiefly insisted on the Method of* Huygens, *which I was absolutely re-*

A
solved

solved to reject, as not seeming to me to be the genuine and natural way of coming at the Solution of Problems of this kind. However, had I allowed my self a little more time to consider it, I had certainly done the Justice to its Author, to have owned that he had not only illustrated Huygens's *Method by a great variety of well chosen Examples, but that he had allied to it several curious things of his own Invention.*

Tho' I have not followed Mr. Huygens *in his Method of Solution, 'tis with very great pleasure that I acknowledge the Obligations I have to him; his Book having Settled in my Mind the first Notions of this Doctrine, and taught me to argue about it with certainty.*

I had said in my Specimen, that Mr. Huygens *was the first who had Published the Rules of this Calculation, intending thereby to do justice to that great Man; but what I then said was misinterpreted, as if I had designed to wrong some Persons who had considered this matter before him, and a passage was cited against me out of* Huygens's *Preface, in which he saith,* Sciendum vero quod jam pridem, inter Præstantissimos totâ Galliâ Geometras, Calculus hic fuerit agitatus; ne quis indebitam mihi primæ Inventionis gloriam hac in re tribuat. *But what follows immediately after, had it been minded, might have cleared me from any Suspicion of injustice. The words are these* Cæterum illi difficillimis quibusque Quæstionibus se invicem exercere Soliti, methodum Suam quisque occultam retinuere, adeo ut a primis elementis hanc materiam evolvere mihi necesse fuerit. *By which it appears, that tho'* Mr. Huygens *was not the first who had applied himself to those sorts of Questions, he was nevertheless the first who had published Rules for their Solution; which is all that I affirmed.*

Since the printing of my Specimen, Mr. de Monmort, *Author of the* Analyse des jeux de Hazard, *Published a Second Edition of that Book, in which he has particularly given many proofs of his singular Genius, and extraordinary Capacity; which Testimony I give both to Truth, and to the Friendship with which he is pleased to Honour me.*

Such a Tract as this is may be useful to several ends; the first of which is, that there being in the World several inquisitive Persons, who are desirous to know what foundation they

go

go upon, when they engage in Play, whether from a motive of Gain, or barely Diversion, they may, by the help of this or the like Tract, gratifie their curiosity, either by taking the pains to understand what is here Demonstrated, or else making use of the conclusions, and taking it for granted that the Demonstrations are right.

Another use to be made of this Doctrine of Chances is, that it may serve in Conjunction with the other parts of the Mathematicks, as a fit introduction to the Art of Reasoning; it being known by experience that nothing can contribute more to the attaining of that Art, than the consideration of a long Train of Consequences, rightly deduced from undoubted Principles, of which this Book affords many Examples. To this may be added, that some of the Problems about Chance having a great appearance of Simplicity, the Mind is easily drawn into a belief, that their Solution may be attained by the meer Strength of natural good Sence; which generally proving otherwise, and the Mistakes occasioned thereby being not unfrequent, 'tis presumed that a Book of this Kind, which teaches to distinguish Truth from what seems so nearly to resemble it, will be look'd upon as a help to good Reasoning.

Among the several Mistakes that are committed about Chance, one of the most common and least suspected, is that which relates to Lotterys. Thus, supposing a Lottery wherein the proportion of the Blanks to the Prizes is as five to one; 'tis very natural to conclude that therefore five Tickets are requisite for the Chance of a Prize; and yet it may be proved Demonstratively, that four Tickets are more then sufficient for that purpose, which will be confirmed by often repeated Experience. In the like manner, supposing a Lottery wherein the proportion of the Blanks to the Prizes is as thirty nine to One, (such as was the Lottery of 1710) it may be proved, that in twenty eight Tickets, a Prize is as likely to be taken as not; which tho' it may seem to contradict the common Notions, is nevertheless grounded upon infallible Demonstration.

When the Play of the Royal Oak was in use, some Persons who lost considerably by it, had their Losses chiefly occasioned by an Argument of which they could not perceive the Fallacy. The Odds against any particular Point of the Ball were one and Thirty to One, which intituled the Adventurers, in case they were winners, to have thirty two Stakes returned, including their own; instead of which they having but eight and Twenty, it was very plain that on the Single account of the disadvantage of the Play, they lost one eighth part of all the Money they play'd for. But the Master of the Ball maintained t'at they had no reason to complain; since he would undertake that any particular point of the Ball should come up in two and Twenty Throws; of this he would offer to lay a Wager, and actually laid it when required. The seeming contradiction between the Odds of one and thirty to One, and Twenty two Throws for any Chance to come up, so perplexed the Adventurers, that they begun to think the Advantage was on their side; for which reason they play'd on and continued to lose.

The Doctrine of Chances may likewise be a help to cure a Kind of Superstition, which has been of long standing in the World, viz. that there is in Play such a thing as Luck, good or bad. I own there are a great many judicious people, who without any other Assistance than that of their own reason, are satisfied, that the Notion of Luck is meerly Chimerical; yet I conceive that the ground they have to look upon it as such, may still be farther inforced from some of the following Considerations.

If by saying that a Man has had good Luck, nothing more was meant than that he has been generally a Gainer at play, the Expression might be allowed as very proper in a short way of speaking: But if the Word good Luck be understood to signifie a certain predominant quality, so inherent in a Man, that he must win whenever he Plays, or at least win oftner than lose, it may be denied that there is any such thing in nature.

The Asserters of Luck are very sure from their own Experience, that at some times they have been very Lucky, and that at other times they have had a prodigious run of ill Luck against them, which whilst it continued obliged them to be very cautious in engaging with the fortunate; but how Chance should produce those extraordinary Events, is what they cannot conceive; They would be glad for Instance to be Satisfied, how they could lose Fifteen Games together at Piquet, if ill Luck had not strangely prevailed against them. But if they will be pleased to consider the Rules delivered in this Book, they will see that tho' the Odds against their losing so many times together be very

great

great, viz. 32767 to 1, yet that the Possibility of it is not destroy'd by the greatness of the Odds, there being One Chance in 32768 that it may so happen, from whence it follows, that it was still possible to come to pass without the Intervention of what they call ill Luck.

Again, This Accident of losing Fifteen times together at Piquet, is no more to be imputed to ill Luck, than the Winning with one single Ticket the Highest Prize, in a Lottery of 32768 Tickets, is to be imputed to good Luck, since the Chances in both Cases are perfectly equal. But if it be said that Luck has been concerned in this latter Case, the Answer will be easy; for let us suppose Luck not existing, or at least let us suppose its Influence to be suspended, yet the Highest Prize must fall into some Hand or other, not by Luck, (for by the Hypothesis that has been laid aside) but from the meer Necessity of its falling somewhere.

Those who contend for Luck, may, if they please, alledge other Cases at Play, much more unlikely to happen than the Winning or Losing Fifteen Games together, yet still their Opinion will never receive any Addition of Strength from such Suppositions : For, by the Rules of Chance, a time may be computed, in which those Cases may as probably happen as not ; nay, not only so, but a time may be computed in which there may be any proportion of Odds for their so happening.

But supposing that Gain and Loss were so fluctuating, as always to be distributed equally, whereby Luck would certainly be annihilated ; would it be reasonable in this Case to attribute the Events of Play to Chance alone? I think, on the contrary, it would be quite otherwise, for then there would be more reason to suspect that some unaccountable Fatality did Rule in it : Thus, If two Persons play at Cross and Pile, and Chance alone be suppos'd to be concern'd in regulating the fall of the Piece, is it probable that there should be an Equality of Heads and Crosses ? It is Five to Three that in four times there will be an inequality; 'tis Eleven to Five in six, 93 to 35 in Eight, and about 12 to 1 in a hundred times : Wherefore Chance alone by its Nature constitutes the Inequalities of Play, and there is no need to have recourse to Luck to explain them.

Further, The same Arguments which explode the Notion of Luck, may, on the other side, be useful in some Cases to establish a due comparison between Chance and Design : We may imagine Chance and Design to be as it were in Competition with each other, for the production

a

duction

duction of some sorts of Events, and may calculate what Probability there is, that those Events should be rather owing to one than to the other. To give a familiar Instance of this, Let us suppose that two Packs of Piquet Cards being sent for, it should be perceived that there is, from Top to Bottom, the same Disposition of the Cards in both Packs; Let us likewise suppose that, some doubt arising about this Disposition of the Cards, it should be questioned whether it ought to be attributed to Chance, or to the Maker's Design: In this case the Doctrine of Combination decides the Question, since it may be proved by its Rules, that there are the Odds of above 263130 3 Millions of Millions of Millions of Millions to One, that the Cards were designedly set in the Order in which they were found.

From this last Consideration we may learn, in many Cases, how to distinguish the Events which are the effect of Chance, from those which are produc'd by Design: The very Doctrine that finds Chance where it really is, being able to prove by a gradual Increase of Probability, till it arrive at Demonstration, that where Uniformity, Order and Constancy reside, there also reside Choice and Design.

Lastly, One of the Principal Uses to which this Doctrine of Chances may be apply'd, is the discovering of some Truths, which cannot fail of pleasing the Mind, by their Generality and Simplicity; the Admirable Connexion of its Consequences will increase the Pleasure of the Discovery; and the seeming Paradoxes wherewith it abounds, will afford very great matter of Surprize and Entertainment to the Inquisitive. A very remarkable Instance of this nature may be seen in the prodigious Advantage which the repetition of Odds will amount to; Thus, Supposing I play with an Adversary who allows me the Odds of 43 to 40, and agrees with me to play till 100 Stakes are won or lost on either side, on condition that I give him an Equivalent for the Gain I am intitled to by the Advantage of my Odds; the Question is what Equivalent I am to give him, on supposition we play a Guinea a Stake: The Answer is 99 Guineas and above 18 Shillings, which will seem almost incredible, considering the smalness of the Odds of 43 to 40. Now let the Odds be in any Proportion given, and let the Number of Stakes to be played for be never so great, yet one General Conclusion will include all the possible Cases, and the application of it to Numbers may be wrought in less than a Minutes time.

I have explain'd, in my Introduction to the following Treatise, the chief Rules on which the whole Art of Chances depends; I have

done

done it in the plainest manner that I could think of, to the end it might be (as much as possible) of General Use. I flatter my self that those who are acquainted with Arithmetical Operations, will, by the help of the Introduction alone, be able to solve a great Variety of Questions depending on Chance: I wish, for the Sake of some Gentlemen who have been pleased to subscribe to the printing of my Book, that I could every where have been as plain as in the Introduction; but this was hardly practicable, the Invention of the greatest part of the Rules being intirely owing to Algebra; yet I have, as much as possible, endeavour'd to deduce from the Algebraical Calculation several practical Rules, the Truth of which may be depended upon, and which may be very useful to those who have contented themselves to learn only common Arithmetick.

'Tis for the Sake of those Gentlemen that I have enlarged my first Design, which was to have laid down such Precepts only as might be sufficient to deduce the Solution of any difficult Problem relating to my Subject: And for this reason I have (towards the latter end of the Book) given the Solution, in Words at length, of some easy Problems, which might else have been made Corollaries or Consequences of the Rules before deliver'd: The single Difficulty which may occur from Pag. 155 to the end, being only an Algebraical Calculation belonging to the 49th Problem, to explain which fully would have required too much room.

On this Occasion, I must take notice to such of my Readers as are well vers'd in Vulgar Arithmetick, that it would not be difficult for them to make themselves Masters, not only of all the Practical Rules in this Book, but also of more useful Discoveries, if they would take the small Pains of being acquainted with the bare Notation of Algebra, which might be done in the hundredth part of the Time that is spent in learning to read Short-hand.

One of the Principal Methods I have made use of in the following Treatise, has been the Doctrine of Combinations, taken in a Sence somewhat more extensive, than as it is commonly understood. The Notion of Combinations being so well fitted to the Calculation of Chance, that it naturally enters the Mind whenever any Attempt is made towards the Solution of any Problem of that kind. It was this which led me in course to the Consideration of the Degrees of Skill in the Adventurers at Play, and I have made use of it in most parts of this Book, as one of the Data that enter the Question; it being so far from perplexing the Calculation, that on

the

the contrary it is rather a Help and an Ornament to it: It is true, that this Degree of Skill is not to be known any other way than from Observation; but if the same Observation constantly recur, 'tis strongly to be presumed that a near Estimation of it may be made: However, to make the Calculation more precise, and to avoid causing any needless Scruples to those who love Geometrical Exactness, it will be easy, in the room of the Word Skill, to substitute a Greater or Less Proportion of Chances among the Adventurers, so as each of them may be said to have a certain Number of Chances to win one single Game.

The General Theorem invented by Sir Isaac Newton, for raising a Binomial to any Power given, facilitates infinitely the Method of Combinations, representing in one View the Combination of all the Chances, that can happen in any given Number of Times. 'Tis by the help of that Theorem, joined with some other Methods, that I have been able to find practical Rules for the solving a great Variety of difficult Questions, and to reduce the Difficulty to a single Arithmetical Multiplication, whereof several Instances may be seen in the 21st Page of this Book.

Another Method I have made use of is that of Infinite Series, which in many cases will solve the Problems of Chance more naturally than Combinations. To give the Reader a Notion of this, we may suppose two Men at Play throwing a Die, each in their Turns, and that he be to be reputed the Winner who shall first throw an Ace: It is plain, that the Solution of this Problem cannot so properly be reduced to Combinations, which serve chiefly to determine the proportion of Chances between the Gamesters, without any regard to the Priority of Play. 'Tis convenient therefore to have recourse to some other Method, such as the following. Let us suppose that the first Man, being willing to Compound with his Adversary for the Advantage he is intitled to from his first Throw, should ask him what Consideration he would allow to yield it to him; it may naturally be supposed that the Answer would be one Sixth part of the Stake, there being but Five to One against him, and that this Allowance would be thought a just Equivalent for yielding his Throw: Let us likewise suppose the Second Man to require in his Turn to have one Sixth part of the remaining Stake for the Consideration of his Throw; which being granted, and the first Man's Right returning in course, he may claim again one Sixth part of the Remainder, and so on alternately, till the whole Stake be exhausted:

But

PREFACE.

But this not being to be done till after an infinite number of Shares be thus taken on both Sides, it belongs to the Method of Infinite Series to assign to each Man what proportion of the Stake he ought to take at first, so as to answer exactly that fictitious Division of the Stake in infinitum; by means of which it will be found, that the Stake ought to be Divided between the contending Parties into two parts, respectively proportional to the two Numbers 6 and 5. By the like Method it would be found that if there were Three or more Adventurers playing on the conditions above described, each Man, according to the Situation he is in with respect to Priority of Play, might take as his Due such part of the Stake, as is expressible by the corresponding Term of the proportion of 6 to 5, continued to so many Terms as there are Gamesters; which in the case of Three Gamesters, for Instance, would be the Numbers 6, 5 and $4\frac{1}{6}$, or their Proportionals 36, 30, and 25.

Another Advantage of the Method of Infinite Series is, that every Term of the Series includes some particular Circumstance wherein the Gamesters may be found, which the other Methods do not; and that a few of its Steps are sufficient to discover the Law of its Process. The only Difficulty which attends this Method, being that of Summing up so many of its Terms as are requisite for the Solution of the Problem proposed: But it will be found by experience, that in the Series resulting from the consideration of most Cases relating to Chance, the Terms of it will either constitute a Geometric Progression, which by the Known Methods is easily Summable; or else some other sort of Progression, whose nature consists in this, that every Term of it has to a determinate number of the preceding Terms, each being taken in order, some constant relation; in which case I have contrived some easie Theorems, not only for finding the Law of that relation, but also for finding the Sums required; as may be seen in several places of this Book, but particularly from page 127 to page 134. I hope the Reader will excuse my not giving the Demonstrations of some few things relating to this Subject, especially of the two Theorems contained in page 134 and 154, and of the Method of Approximation contained in page 149 and 150; whereby the Duration of Play is easily determined with the help of a Table of Natural Sines: Those Demonstrations are omitted purposely to give an occasion to the Reader to exercise his own Ingenuity. In the mean Time, I have deposited them with the Royal Society, in order to be Published when it shall be thought requisite.   b   A

A Third Advantage of the Method of Infinite Series is, that the Solutions derived from it have a certain Generality and Elegancy, which scarce any other Method can attain to; those Methods being always perplexed with Various unknown Quantities, and the Solutions obtained by them terminating commonly in particular Cases.

There are other Sorts of Series, which tho' not properly infinite, yet are called Series, from the Regularity of the Terms whereof they are composed; those Terms following one another with a certain uniformity, which is always to be defined. Of this nature is the Theorem given by Sir Isaac Newton, in the Fifth Lemma of the third Book of his Principles, for drawing a Curve through any given number of Points; of which the Demonstration, as well as of other things belonging to the same Subject, may be deduced from the first Proposition of his Methodus Differentialis, printed with some other of his Tracts, by the care of my Intimate Friend, and very skilful Mathematician, Mr. W. Jones. The abovementioned Theorem being very useful in Summing up any number of Terms whose last Differences are equal (Such as are the Numbers called Triangular, Pyramidal, &c. the Squares, the Cubes, or other Powers of Numbers in Arithmetic Progression) I have shewn in many places of this Book how it might be applicable to these Cases. I hope it will not be taken amiss that I have ascribed the Invention of it to its proper Author, tho' 'tis possible some Persons may have found something like it by their own Sagacity.

After having dwelt some time upon Various Questions depending on the general Principle of Combinations, as laid down in my Introduction, and upon some others depending on the Method of Infinite Series, I proceed to treat of the Method of Combinations properly so called, which I shew to be easily deducible from that more general Principle which hath been before explained: Where it may be observed, that altho' the Cases it is applyed to are particular, yet the way of reasoning, and the consequences derived from it, are general; that Method of Arguing about generals by particular Examples, being in my opinion very convenient for easing the Reader's Imagination.

Having explained the common Rules of Combination, and given a Theorem which may be of use for the Solution of some Problems relating to that Subject, I lay down a new Theorem, which is properly a contraction of the former, whereby several Questions of Chance are resolved with wonderful ease, tho' the Solution might seem at first sight to be of insuperable difficulty.

It is by the Help of that Theorem so contracted, that I have been able to give a compleat Solution of the Problems of Pharaon *and* Bassete, *which was never done till now: I own that some great Mathematicians have before me taken the pains of calculating the Advantage of the* Banker, *in any circumstance either of Cards remaining in his Hands, or of any number of times that the Card of the* Ponte *is contained in the Stock: But still the curiosity of the Inquisitive remained unsatisfied; The Chief Question, and by much the most difficult, concerning* Pharaon *or* Bassete, *being what it is that the Banker gets* per Cent *of all the Money adventured at those Games, which now I can certainly answer is very near* Three per Cent *at* Pharaon, *and* Three fourths per Cent *at* Bassete, *as may be seen in my* xxiii *Problem, where the precise Advantage is calculated.*

In the 24th *and* 25th *Problems, I explain a new sort of* Algebra, *whereby some Questions relating to Combinations are solved by so easy a Process, that their solution is made in some measure an immediate consequence of the Method of Notation. I will not pretend to say that this new* Algebra *is absolutely necessary to the Solving of those Questions which I make to depend on it, since it appears by Mr.* De Monmort's Book, *that both he and Mr.* Nicholas Bernoully *have solved, by another Method, many of the cases therein proposed: But I hope I shall not be thought guilty of too much Confidence, if I assure the Reader, that the Method I have followed has a degree of Simplicity, not to say of Generality, which will hardly be attained by any other Steps than by those I have taken.*

The 29th *Problem, proposed to me, amongst some others, by the Honourable Mr.* Francis Robartes, *I had solved in my Tract* De mensura Sortis; *It relates, as well as the* 24th *and* 25th, *to the Method of Combinations, and is made to depend on the same Principle; When I began for the first time to attempt its Solution, I had nothing else to guide me but the common Rules of Combinations, such as they had been delivered by Dr.* Wallis *and others; which when I endeavoured to apply, I was Surprized to find that my calculation swelled by degrees to an Intolerable bulk: For this reason I was forced to turn my Views Another way, and to try whether the solution I was seeking for might not be deduced from some easier considerations; whereupon I happily fell upon the Method I have been mentioning, which as it led me to a very great Simplicity in the Solution, so I look upon it to be an Improvement made to the Method of Combinations.*

The 30th Problem is the reverse of the preceding; It contains a very remarkable Method of Solution, the Artifice of which consists in changing an Arithmetic Progression of Numbers into a Geometric one; this being always to be done when the Numbers are large, and their Intervals small. I freely acknowledge that I have been indebted long ago for this useful Idea, to my much respected Friend, That Excellent Mathematician Doctor Halley, Secretary to the Royal Society, whom I have seen practice the thing on an other occasion: For this and other Instructive Notions readily imparted to me, during an uninterrupted Friendship of five and Twenty years, I return him my very hearty Thanks.

The 32d Problem, having in it a Mixture of the two Methods of Combinations and Infinite Series, may be proposed for a pattern of Solution, in some of the most difficult cases that may occurr in the Subject of Chance, and on this occasion I must do that Justice to Mr. Nicholas Bernoully, the Worthy Professour of Mathematics at Padua, to own he had sent me the Solution of this Problem before mine was Published; which I had no sooner received, but I communicated it to the Royal Society, and represented it as a Performance highly to be commended: Whereupon the Society order'd that his Solution should be Printed; which was accordingly done some time after in the Philosophical Transactions, Numb. 341. where mine was also inserted.

The Problems which follow relate chiefly to the Duration of Play, or to the Method of determining what number of Games may probably be played out by two Adversaries, before a certain number of Stakes agreed on between them be won or lost on either side. This Subject affording a very great Variety of Curious Questions, of which every one has a degree of Difficulty peculiar to it self, I thought it necessary to divide it into several distinct Problems, and to illustrate their Solution with proper Examples.

Tho' these Questions may at first fight seem to have a very great degree of difficulty, yet I have some reason to believe, that the Steps I have taken to come at their Solution, will easily be followed by those who have a competent skill in Algebra, and that the chief Method of proceeding therein will be understood by those who are barely acquainted with the Elements of that Art.

When I first began to attempt the general Solution of the Problem concerning the Duration of Play, there was nothing extant that could give me any light into that Subject; for altho' Mr.

de

de Monmort, *in the first Edition of his Book, gives the Solution
of this Problem, as limited to three Stakes to be won or lost, and
farther limited by the Suppofition of an Equality of Skill between
the Adventurers; yet he having given no Demonftration of his
Solution, and the Demonftration when I fearced being of very little ufe towards obtaining the general Solution of the Problem, I
was forced to try what my own Enquiry would lead me to, which
having been attended with Succefs, the refult of what I found was
afterwards published in my* Specimen *before mentioned.*

All the Problems which in my Specimen *related to the Duration of Play, have been kept entire in the following Treatife; but
the Method of Solution has received fome Improvements by the
new Difcoveries I have made concerning the Nature of thofe Series
which refult from the Confideration of the Subject; however, the
Principles of that Method having been laid down in my* Specimen
*I had nothing now to do, but to draw the Confequences that were
naturally deducible from them.*

Mr. de Monmort, *and Mr.* Nicholas Bernoully, *have each of
them feparately given the Solution of my* xxxixth *Problem, in a
Method differing from mine, as may be feen in Mr.* de Monmort's
*fecond Edition of his Book. Their Solutions, which in the main
agree together, and vary little more than in the form of Expreffion,
are extreamly beautiful; for which reafon I thought the Reader
would be well pleafed to fee their Method explained by me, in fuch a
manner as might be apprehended by thofe who are not fo well verfed in
the nature of Symbols: In which matter I have taken fome Pains,
thereby to teftify to the World the juft Value I have for their
Performance.*

The 43d *Problem having been propofed to me by Mr.* Thomas
Woodcock, *a Gentleman whom I infinitely refpect, I attempted its
Solution with a very great defire of obtaining it; and having had
the good Fortune to fucceed in it, I returned him the Solution a few
Days after he was pleafed to propofe it. This Problem is in my
Opinion one of the moft curious that can be propos'd on this Subject;
its Solution containing the Method of determining, not only that
Advantage which refults from a Superiority of Chance, in a Play
confined to a certain number of Stakes to be won or loft by either
Party, but alfo that which may refult from an unequality of Stakes;
and even compares thofe two Advantages together, when the Odds of
Chance being on one fide, the Odds of Money are on the other.*

c

Before

Before I make an end of this Discourse, I think my self obliged to take Notice, that some Years after my Specimen was printed, there came out a Tract upon the Subject of Chance, being a Posthumous Work of Mr. James Bernoully, *wherein the Author has shewn a great deal of Skill and Judgment, and perfectly answered the Character and great Reputation he hath so justly obtained. I wish I were capable of carrying on a Project he had begun, of applying the Doctrine of Chances to* Oeconomical *and* Political *Uses, to which I have been invited, together with Mr* de Monmort, *by Mr.* Nicholas Bernoully : *I heartily thank that Gentleman for the good Opinion he has of me ; but I willingly resign my share of that Task into better Hands, wishing that either he himself would prosecute that Design, he having formerly published some successful Essays of that Kind, or that his Uncle, Mr.* John Bernoully, *Brother to the Deceased, could be prevailed upon to bestow some of his Thoughts upon it ; he being known to be perfectly well qualified in all Respects for such an Undertaking.*

Due Care having been taken to avoid the *Errata* of the Press, we hope there are no other than these two,

V I Z.

Pag. 35. Lin. 35. for *n* — 3 read *n* — 1.
Pag. 36. Lin. 2. for *n* — 3 read *n* — 2.

The

UNIV. OF
...ORNL

The DOCTRINE

OF

CHANCES.

INTRODUCTION.

HE Probability of an Event is greater, or lefs, according to the number of Chances by which it may Happen, compar'd with the number of all the Chances, by which it may either Happen or Fail.

Thus, If an Event has 3 Chances to Happen, and 2 to Fail; the Probability of its Happening may be eftimated to be $\frac{3}{5}$, and the Probability of its Failing $\frac{2}{5}$.

Therefore, if the Probability of Happening and Failing are added together, the Sum will always be equal to Unity.

If

If the Probabilities of Happening and Failing are unequal, there is what is commonly call'd Odds for, or against, the Happening or Failing; which Odds are proportional to the number of Chances for Happening or Failing.

The Expectation of obtaining any Thing, is estimated by the Value of that Thing multiplied by the Probability of obtaining it.

Thus, Supposing that A and B Play together; that A has deposited 5 l, and B 3 l; that the number of Chances which A has to win is 4, and that the number of Chances which B has to win is 2: Since the whole Sum deposited is 8, and that the Probability which A has of getting it, is $\frac{4}{6}$; it follows, that the Expectation of A upon the whole Sum deposited will be $\frac{8}{1} \times \frac{4}{6} = 5\frac{1}{3}$; and for the same reason, the Expectation of B will be $\frac{8}{1} \times \frac{2}{6} = 2\frac{2}{3}$.

The Risk of losing any Thing, is estimated by the Value of that Thing multiplied by the Probability of losing it.

If from the respective Expectations, which the Gamesters have upon the whole Sum deposited; the particular Sums they deposit, that is their own Stakes, be subtracted, there will remain the Gain, if the difference is positive, or the Loss, if the difference is negative.

Thus, If from $5\frac{1}{3}$ the Expectation of A, 5 which is his own Stake be subtracted, there will remain $\frac{1}{3}$ for his Gain; likewise if from $2\frac{1}{3}$ the Expectation of B, 3 which is his own Stake be subtracted, there will remain $-\frac{1}{3}$ for his Gain, or $\frac{1}{3}$ for his Loss.

Again, If from the respective Expectations, which either Gamester has upon the Sum deposited by his Adversary, the Risk of losing what he himself deposits, be subtracted, there will likewise remain his Gain or Loss.

Thus, In the preceding Case, the Stake of B being 3, and the Probability which A has of winning it being $\frac{4}{6}$, the Expectation of A upon that Stake is $\frac{3}{1} \times \frac{4}{6} = \frac{12}{6} = 2$. Moreover the Stake of A being 5, and the Probability of losing it being $\frac{2}{6}$, the Risk which A runs of losing his own Stake is $\frac{5}{1} \times \frac{2}{6} = \frac{10}{6} = 1\frac{2}{3}$. Therefore, if from the

Ex-

Expectation 2, the Risk $1\frac{2}{3}$, be subtracted, there will remain $\frac{1}{3}$, as before, for the Gain of A; and by the same way of arguing, the Loss of B will be found to be $\frac{1}{3}$.

N. B. Tho' the Gain of one is the Loss of the other, yet it will be convenient to look for them severally, that one Operation may be a Proof of the other.

If there is a certain number of Chances by which the possession of a Sum can be secur'd; and also a certain number of Chances by which it may be lost; that Sum may be Insured for that part of it, which shall be to the whole, as the number of Chances there is to lose it, to the number of all the Chances.

Thus, If there are 19 Chances to secure the possession of 1000 *l*, and 1 Chance to lose it, the Insurance Money may be found by this Proportion.

As 20 is to 1, so is 1000 to 50; therefore 50 is the Sum that ought to be given, in this Case, to Insure 1000.

If two Events have no dependence on each other, so that p be the number of Chances by which the first may Happen, and q the number of Chances by which it may Fail; and likewise that r be the number of Chances by which the second may Happen, and s the number of Chances by which it may Fail: Multiply $p + q$ by $r + s$, and the Product $pr + qr + ps + qs$ will contain all the Chances, by which the Happening, or Failing of the Events may be varied amongst one another.

Therefore, If A and B Play together, on condition that if both Events Happen, A shall win, and B lose; the Odds that A shall be the winner, are as pr to $qr + ps + qs$; for the only Term in which both p and r occur is pr; therefore the Probability of A's winning is $\frac{pr}{pr + qr + ps + qs}$.

But if A holds that either one or the other will Happen; the Odds of A's winning are as $pr + qr + ps$ to qs; for some of the Chances that are favourable to A, occur in every one of the Terms pr, qr, ps.

Again, If A holds that the first will Happen, and the second Fail; the Odds are as ps to $pr + qr + qs$.

From

From what has been faid, it follows, that if a Fraction ex-preffes the Probability of an Event, and another Fraction the Probability of another Event, and thofe two Events are in-dependent; the Probability that both thofe Events will Hap-pen, will be the Product of thofe two Fractions.

Thus, Suppofe I have two Wagers depending, in the firft of which I have 3 to 2 the beft of the Lay, and in the fecond 7 to 4, what is the Probability I win both Wagers?

The Probability of winning the firft is $\frac{3}{5}$, that is the num-ber of Chances I have to win, divided by the number of all the Chances; the Probability of winning the fecond is $\frac{7}{11}$: Therefore multiplying thefe two Fractions together, the Pro-duct will be $\frac{21}{55}$, which is the Probability of winning both Wagers. Now this Fraction being fubtracted from 1, the remainder is $\frac{34}{55}$, which is the Probability I do not win both Wagers: Therefore the Odds againft me are 34 to 21.

2° If I would know what the Probability is of winning the firft, and lofing the fecond, I argue thus; The Probability of winning the firft is $\frac{3}{5}$, the Probability of lofing the fe-cond is $\frac{4}{11}$: Therefore multiplying $\frac{3}{5}$ by $\frac{4}{11}$, the Product $\frac{12}{55}$ will be the Probability of my winning the firft, and lofing the fecond; which being fubtracted from 1, there will remain $\frac{43}{55}$, which is the Probability I do not win the firft, and at the fame time lofe the fecond.

3° If I would know what the Probability is of winning the fecond, and at the fame time lofing the firft; I fay thus, the Probability of winning the fecond is $\frac{7}{11}$, the Probability of lofing the firft is $\frac{2}{5}$. Therefore multiplying thefe two Fractions together, the Product $\frac{14}{55}$ is the Probability I win the fecond, and alfo lofe the firft.

4° If I would know what the Probability is of lofing both Wagers; I fay, the Probability of lofing the firft is $\frac{2}{5}$, and the Probability of lofing the fecond $\frac{4}{11}$; therefore the Proba-bility of lofing them both is $\frac{8}{55}$, which being fubtracted from 1, there remains $\frac{47}{55}$; therefore the Odds againft lofing both Wagers is 47 to 8.

                                                              This

This way of reafoning is plain, and is of very great extent, being applicable to the Happening or Failing of as many Events as may fall under confideration. Thus, if I would know what the Probability is of miffing an Ace 4 times together with a common Die, I confider the miffing of the Ace 4 times, as the Failing of 4 different Events; now the Probability of miffing the firft is $\frac{5}{6}$, the Probability of miffing the fecond is alfo $\frac{5}{6}$, the third $\frac{5}{6}$, the fourth $\frac{5}{6}$; therefore the Probability of miffing the Ace 4 times together is $\frac{5}{6} \times \frac{5}{6} \times \frac{5}{6} \times \frac{5}{6} = \frac{625}{1296}$; which being fubtracted from 1, there will remain $\frac{671}{1296}$ for the Probability of throwing it once or oftner in 4 times; therefore the Odds of throwing an Ace in 4 times is 671 to 625.

But if the flinging of an Ace was undertaken in 3 times, 'tis plain that the Probability of miffing it 3 times would be $\frac{5}{6} \times \frac{5}{6} \times \frac{5}{6} = \frac{125}{216}$, which being fubtracted from 1, there will remain $\frac{91}{216}$ for the Probability of throwing it once or oftner in 3 times; therefore the Odds againft throwing it in 3 times are 125 to 91.

Again, fuppofe we wou'd know the Probability of throwing an Ace once in 4 throws and no more: Since the Probability of throwing it the firft time is $\frac{1}{6}$, and the Probability of miffing it the other three times is $\frac{5}{6} \times \frac{5}{6} \times \frac{5}{6}$, it follows that the Probability of throwing it the firft time, and miffing it afterwards three times fucceffively, is $\frac{1}{6} \times \frac{5}{6} \times \frac{5}{6} \times \frac{5}{6} = \frac{125}{1296}$; but becaufe it is poffible to hit it in every throw as well as the firft, it follows, that the Probability of throwing it once in 4 throws, and miffing the other three times, is $\frac{4 \times 125}{1296} = \frac{500}{1296}$; which being fubtracted from 1, there will remain $\frac{796}{1296}$ for the Probability of not throwing it once and no more in 4 times; therefore if one undertakes to throw an Ace once and no more in 4 times, he has 500 to 796 the worft of the Lay, or 5 to 8 very near.

Suppofe two Events are fuch, that one of them has twice as many Chances to come up as the other; what is the Probability that the Event which has the greater number of Chances to come up, does not Happen twice before the other Happens once; which is the Cafe of flinging Seven with two Dice, be-

fore

fore four once ; fince the number of Chances are as 2 to 1, the Probability of the firft Happening before the fecond is $\frac{2}{3}$, but the Probability of its Happening twice before it, is but $\frac{2}{3} \times \frac{2}{3}$ or $\frac{4}{9}$; therefore 'tis 5 to 4, Seven does not come up twice, before Four once.

But if it was demanded what muft be the proportion of the Facilities of the coming up of two Events, to make that which has the moft Chances, to come up twice, before the other comes up once ; the anfwer is 12 to 5 very near, (and this proportion may be determined yet with greater exact-nefs) for if the proportion of the Chances is 12 to 5, it follows, that the Probability of throwing the firft before the fecond is $\frac{12}{17}$, and the Probability of throwing it twice, is $\frac{12}{17} \times \frac{12}{17}$ or $\frac{144}{289}$; therefore the Probability of not doing it is $\frac{145}{289}$; therefore the Odds againft it are as 145 to 144, which comes very near a proportion of Equality.

What we have faid hitherto concerning two or more E-vents, relates only to thofe which have no dependency on each other ; as for thofe that have a dependency, the manner of arguing about them will be a little alter'd : But to know in what the nature of this dependency confifts, I fhall propofe the two following eafy Problems.

Suppofe there is a heap of 13 Cards of one colour, and a-nother heap of 13 Cards of another colour ; what is the Pro-bability, that taking one Card at a venture out of each heap, I fhall take out the two Aces ?

The Probability of taking the Ace out of the firft heap is $\frac{1}{13}$, the Probability of taking the Ace out of the fecond is alfo $\frac{1}{13}$; therefore the Probability of taking out both Aces is $\frac{1}{13} \times \frac{1}{13}$ $= \frac{1}{169}$, which being fubtracted from 1, there will remain $\frac{168}{169}$, therefore the Odds againft me are 168 to 1.

But fuppofe that out of one fingle heap of 13 Cards of one co-lour, I fhould undertake to take out, firft the Ace, fecondly the Two ; tho' the Probability of taking out the Ace be $\frac{1}{13}$, and the Probability of taking out the Two be likewife $\frac{1}{13}$, yet the Ace being fuppofed as taken out already, there will remain only 12 Cards in the heap, which will make the Probability

of

of taking out the Two to be $\frac{1}{12}$, therefore the Probability of taking out the Ace, and then the Two, will be $\frac{1}{11} \times \frac{1}{12}$. And upon this way of reasoning may the whole Doctrine of Combinations be grounded, as will be shewn in its place.

It is plain that in this last Question, the two Events proposed have on each other a dependency of Order, which dependency consists in this, that one of the Events being supposed as having Happened, the Probability of the other's Happening is thereby alter'd; whereas in the first Question, the taking of the Ace out of the first heap does not alter the Probability of taking the Ace out of the second; therefore the Independency of Events consists in this, that the Happening of one does not alter the degree of Probability of the other's Happening.

We have seen already how to determine the Probability of the Happening of as many Events as may be assigned, and the Failing of as many others as may be assigned likewise, when those Events are independent: We have seen also how to determine the Happening of two Events, or as many as may be assigned when they are Dependent.

But how to determine in the case of Events dependent, the Happening of as many as may be assigned, and at the same time the Failing of as many as may likewise be assigned, is a disquisition of a higher nature, and will be shewn afterwards.

If the Events in question are n in number, and are such as have the same number a of Chances by which they may Happen, and likewise the same number b of Chances by which they may Fail, raise $a + b$ to the Power n.

And if A and B play together, on condition that if either one or more of the Events in question do Happen, A shall win, and B lose; the Probability of A's winning will be $\frac{\overline{a+b}^n - b^n}{\overline{a+b}^n}$, and that of B's winning will be $\frac{b^n}{\overline{a+b}^n}$; for when $a+b$ is actually raised to the Power n, the only Term in which a does not occur is the last b^n; therefore all the Terms but the last are favourable to A.

Thus, if $n = 3$; raising $a + b$ to the Cube $a^3 + 3aab + 3abb + b^3$. all the Terms but b^3 will be favourable to A; and therefore the

the Probability of *A*'s winning will be $\frac{a^3+3aab+1abb}{a+b^3}$ or

$\frac{\overline{a+b}^3-b^3}{a+b^3}$; and the Probability of *B*'s winning will be $\frac{b^3}{a+b^3}$.

But if *A* and *B* play, on condition that if either two or more of the Events in queſtion do Happen, *A* ſhall win; but in caſe one only Happens or none, *B* ſhall win; the Probabi-

lity of *A*'s winning will be $\frac{\overline{a+b}^n - nab^{n-1} - b^n}{a+b^n}$; for the on-

ly two Terms in which *aa* does not occur are the two laſt,

viz, nab^{n-1} and b^n. And ſo of the reſt.

The

.

The Solution of several sorts of Problems, *deduced from the Rules laid down in the* *Introduction.*

PROBLEM I.

Uppose A *to hold a single Die, and to lay with* B, *that in 8 throws he shall fling Two Aces or more : What is his Probability of winning, or what are the Odds for or against him ?*

SOLUTION.

BEcause there is one single Chance for *A*, and five against him, let *a* be made = 1, and *b* = 5; again because the number of throws is 8, let *n* be made = 8, and the Probability of *A*'s winning will be $\frac{\overline{a+b}^n - b^n - nab^{n-1}}{\overline{a+b}^n} = \frac{663991}{1679616}$ Therefore the Probability of his losing will be $\frac{1015625}{1679616}$ and the Odds against him will be as 1015625 to 663991, or as 3 to 2, very near.

PROBLEM II.

TWO *Men* A *and* B *playing a Set together, in each Game of the Set the number of Chances which* A *has to win is* 3, *and the number of Chances which* B *has to win is* 2 : *Now after some Games are over,* A *wants 4 Games of being up, and* B *6: It is required in this circumstance to determine the Probabilities which either has of winning the Set.*

SOLUTION.

BEcause *A* wants 4 Games of being up, and *B* 6; it follows, that the Set will be ended in 9 Games at most, which is the sum of the Games wanting between

D

them;

them; therefore let $a + b$ be raised to the 9th Power, *viz.*
$a^9 + 9a^8 b + 36a^7 bb + 84a^6 b^3 + 126a^5 b^4 + 126a^4 b^5 + 84a^3 b^6 + 36aab^7 + 9ab^8 + b^9$; take for A all the Terms in which a has 4 or more dimensions, and for B all the Terms in which b has 6 or more dimensions; and the Proportion of the Odds will be, as $a^9 + 9a^8 b + 36a^7 bb + 84a^6 b^3 + 126a^5 b^4 + 126a^4 b^5$, to $84a^3 b^6 + 36aab^7 + 9ab^8 + b^9$. Let now a be expounded by 3, and b by 2; and the Odds that A wins the Set will be found as 1759077 to 194048, or very near as 9 to 1.

And generally, supposing that p and q are the number of Games respectively wanting; raise $a+b$ to the Power $p+q-1$, then take for A, a number of Terms equal to q, and for B, a number of Terms equal to p.

REMARKS.

1. In this Problem, if instead of supposing that the Chances which the Gamesters have each time to get a Game are in the proportion of a to b, we suppose the Skill of the Gamesters to be in that proportion, the Solution of the Problem will be the same: We may compare the Skill of the Gamesters to the number of Chances they have to win. Whether the number of Chances which A and B have of getting a Game, are in a certain Proportion, or whether their Skill be in that proportion, is the same thing.

2. The preceding Problem might be solved without Algebra, by the bare help of the Arithmetical Principles which we have laid down in the Introduction, but the method will be longer: Yet for the sake of those who are not acquainted with Algebraical computation, I shall set down the Method of proceeding in like cases.

In order to which, it is necessary to know, that when a Question seems somewhat difficult, it will be useful to solve at first a Question of the like nature, that has a greater degree of simplicity than the case proposed in the Question given; the Solution of which case being obtained, it will be a step to ascend to a case a little more compounded, till at last the case proposed may be attained to.

Therefore, to begin with the simplest case, we may suppose that A wants 1 Game of being up, and B 2; and that
the

the number of Chances to win a Game are equal; in which cafe the Odds that *A* will be up before *B*, may be determined as follows.

Since *B* wants 2 Games of being up and *A* 1, 'tis plain that *B* muft beat *A* twice together to win; but the Probability of his beating him once is $\frac{1}{2}$, therefore the Probability of his beating him twice together is $\frac{1}{2} \times \frac{1}{2} = \frac{1}{4}$; fubtract $\frac{1}{4}$ from 1, there remains $\frac{3}{4}$, which is the Probability which *A* has of winning once before *B* twice, therefore the Odds are as 3 to 1.

By the fame way of arguing 'twill be found, that if *A* wants 1, and *B* 3, the Odds will be as 7 to 1, and the Probability of winning, $\frac{7}{8}$ and $\frac{1}{8}$ refpectively. If *A* wants 1, and *B* 4, the Odds will be as 15 to 1, &c.

Again, fuppofe *A* wants 2, and B 3, what are the Odds that *A* is up before *B*?

Let the whole Stake depofited between *A* and *B* be 1; now confider that if *B* wins the firft Game, *B* and *A* will have an equality of Chances, in which cafe the Expectation of *B* will be $\frac{1}{2}$; but the Probability of his winning the firft Game is $\frac{1}{2}$, therefore the Expectation of *B* upon the Stake, arifing from the Probability of beating *A* the firft time, will be $\frac{1}{2} \times \frac{1}{2} = \frac{1}{4}$.

But if *B* lofes the firft Game, then he will want 3 of being up, and *A* but 1; in which cafe the Expectation of *B* will be $\frac{1}{8}$, but the Probability of that circumftance is $\frac{1}{2}$, therefore the Expectation of *B* arifing from the Probability of his lofing the firft time is $\frac{1}{2} \times \frac{1}{8} = \frac{1}{16}$.

Therefore the Expectation of *B* upon the Stake 1, will be $\frac{1}{4} + \frac{1}{16} = \frac{5}{16}$, which being fubtracted from 1, there remains $\frac{11}{16}$ for the Expectation of *A*; therefore the Odds are as 11 to 5.

And thus proceeding gradually, it will be eafy to compofe the following Table.

A TABLE.

A TABLE of the ODDS for any number of Games wanting, from 1 to 6.

Games wanting		Games wanting		Games wanting		Games wanting		Games wanting	
1	2	1	3	1	4	1	5	1	6
$\frac{3}{4}$	$\frac{1}{4}$	$\frac{7}{8}$	$\frac{1}{8}$	$\frac{15}{16}$	$\frac{1}{16}$	$\frac{31}{32}$	$\frac{1}{32}$	$\frac{63}{64}$	$\frac{1}{64}$
2	3	2	4	2	5	2	6		
$\frac{11}{16}$	$\frac{5}{16}$	$\frac{26}{32}$	$\frac{6}{32}$	$\frac{57}{64}$	$\frac{7}{64}$	$\frac{120}{128}$	$\frac{8}{128}$		
3	4	3	5	3	6				
$\frac{42}{64}$	$\frac{22}{64}$	$\frac{99}{128}$	$\frac{29}{128}$	$\frac{219}{256}$	$\frac{37}{256}$				
4	5	4	6						
$\frac{163}{256}$	$\frac{93}{256}$	$\frac{382}{512}$	$\frac{130}{512}$						
5	6								
$\frac{638}{1024}$	$\frac{386}{1024}$								

And by the same way of proceeding, it would be eafy to compofe other Tables, for exprefling the Probabilities which *A* and *B* have of winning the Set, when each wants a given number of Games of being up, and when the proportion of the Chances

Chances by which each of them may get a Game is as 2 to 1, or varies at pleasure; but the Algebraic method explained in this Problem answering all that variety, 'tis needless to insist upon it.

PROBLEM III.

IF A and B play with single Bowls, and such be the Skill of A that he knows by Experience he can give B 2 Games out of 3: What is the proportion of their Skill, or what are the Odds that A may get any one Game assigned?

SOLUTION.

LET the proportion of the Odds be as z to 1 : Now since A can give B 2 Games out of 3, therefore A can, upon an equality of Play, undertake to win 3 Games together: Let therefore $z + 1$ be raised to the Cube, viz. $z^3 + 3zz + 3z + 1$; therefore the Probabilities of winning will be, as z^3 to $3zz + 3z + 1$; but these Probabilities are equal, by supposition; therefore $z^3 = 3zz + 3z + 1$, or $2 z^3 = z^3 + 3zz + 3z + 1$. and extracting the Cube Root on both sides, $z\sqrt[3]{2} = z + 1$; therefore $z = \frac{1}{\sqrt[3]{2}-1}$, and consequently the Odds that A may get any one Game assigned are as $\frac{1}{\sqrt[3]{2}-1}$ to 1, or as 1 to $\sqrt[3]{2}-1$, that is in this case as 50 to 13 very near.

PROBLEM IV.

IF A can without advantage or disadvantage give B 1 Game out of 3; what are the Odds that A shall take any one Game assigned? Or what is the proportion of the Chances they have to win any one Game assigned? Or what is the proportion of their Skill?

SOLUTION.

LET the proportion be as z to 1; now since A can give B 1 Game out of 3; therefore A can, upon an equality of Play, undertake to get 3 Games before B gets 2; let there-

B　　fore

fore $z+1$ be raifed to the $4th$ Power, whofe Index 4 is the Sum of the Games wanting between them lefs by 1; this Power will be $z^4 + 4z^3 + 6zz + 4z + 1$; therefore the Probabilities of winning the Set will be as $z^4 + 4z^3$ to $6zz + 4z + 1$: But thefe Probabilities are equal by Hypothefis, fince A and B are fuppofed to play without advantage or difadvantage; therefore $z^4 + 4z^3 = 6zz + 4z + 1$, which Equation being folved, z will be found to be 1.6 very near; wherefore the proportion of the Odds will be as 1.6 to 1, or as 8 to 5.

PROBLEM V.

TO *find in how many Trials an Event will Probably Happen, or how many Trials will be requifite to make it indifferent to lay on its Happening or Failing; fuppofing that* a *is the number of Chances for its Happening in any one Trial, and* b *the number of Chances for its Failing.*

SOLUTION.

LET x be the number of Trials; therefore by what has been already demonftrated in the Introduction $\overline{a+b}^x - b^x = b^x$, or $\overline{a+b}^x = 2b^x$; therefore $x = \frac{Lg\ 2}{Lg:a+b - Lg:b}$.

Moreover, let us reaffume the Equation $\overline{a+b}^x = 2b^x$ and making $a, b :: 1, q$, the Equation will be changed into this $\overline{1+\frac{1}{q}}^x = 2$: let therefore $1 + \frac{1}{q}$ be raifed actually to the Power x by Sir *Ifaac Newton*'s Theorem, and the Equation will be $1 + \frac{x}{q} + \frac{x \times x - 1}{1 \times 2qq} + \frac{x \times x - 1 \times x - 2}{1 \times 2 \times 3 q^3}$ &c. $= 2$. In this Equation, if $q = 1$, then will x be likewife $= 1$; if q be infinite, then will x alfo be infinite. Suppofe q infinite, then the Equation will be reduced to $1 + \frac{x}{q} + \frac{xx}{2qq} + \frac{x^3}{6q^3}$ &c. $= 2$: But the firft part of this Equation is the number whofe Hyperbolic Logarithm is $\frac{x}{q}$, therefore $\frac{x}{q} = $ Log: 2: But the Hyperbolic Logarithm of 2 is 0.693 or nearly 0.7; Wherefore $\frac{x}{q} = 0.7$, and $x = 0.7q$ very near.

Thus we have affigned the very narrow limits within which the Ratio of x to q is comprehended; for it begins with

with Unity, and terminates at laſt in the Ratio of 10 to 7, very near.

But *x* ſoon Converges to the limit 0.7 *q*, ſo that this pro-portion may be aſſumed in all caſes, let the Value of *q* be what it will.

Some uſes of this Propoſition will appear by the following Examples.

EXAMPLE I.

LET it be propoſed to find in how many Throws one may undertake, with an equality of Chance, to fling two Aces with two Dice.

The number of Chances upon two Dice is 36, out of which there is but 1 Chance for two Aces; therefore the number of Chances againſt it is 215: Multiply 35 by 0.7, and the product 24.5 will ſhew that the number of Throws requiſite to that effect will be between 24 and 25.

EXAMPLE II.

TO find in how many Throws of three Dice, one may undertake to fling three Aces.

The number of all the Chances upon 3 Dice is 216, out of which there is but 1 Chance for 3 Aces, and 215 againſt it. Therefore let 215 be multiplied by 0.7, and the product 150.5, will ſhew that the number of Chances requiſite to that effect will be 150, or very near it.

EXAMPLE III.

IN a Lottery whereof the number of Blanks is to the num-ber of Prizes as 39 to 1, (ſuch as was the Lottery of 1710;) To find how many Tickets one muſt take, to make it an equal Chance for one or more Prizes.

Multiply 39 by 0.7, and the product 27.3 will ſhow that the number of Tickets requiſite to that effect will be 27, or 28 at moſt.

Likewiſe, in a Lottery whereof the number of Blanks is to the number of Prizes; as 5 to 1, multiply 5 by 0.7 and the
pro-

product 3.5 will fhow, that there is more than an equality of Chance in 4 Tickets for one or more Prizes, but fomething lefs than an equality in 3.

REMARK.

In a Lottery whereof the Blanks are to the Prizes as 39 to 1, if the number of Tickets in all was but 40, this proportion would be altered, for 20 Tickets would be a fufficient number for the Expectation of the fingle Prize; it being evident that the Prize may be as well among the Tickets which are taken as among thofe that are left behind.

Again, if the number of Tickets was 80, ftill preferving the proportion of 39 Blanks to 1 Prize, and confequently fuppofing 78 Blanks to 2 Prizes, this proportion would ftill be altered: For by the Doctrine of Combinations, whereof we are to treat afterwards, it will appear that the Probability of taking one Prize or both in 20 Tickets would be but $\frac{139}{316}$, and the Probability of taking none would be $\frac{177}{316}$; Wherefore the Odds againft taking any Prize would be as 177 to 139, or very near as 9 to 7.

And by the fame Doctrine of Combinations it will be found that 23 Tickets would not be quite fufficient for the Expectation of a Prize in this Lottery; but that 24 would rather be two many; fo that one might with advantage lay an even Wager of taking a Prize in 24 Tickets.

If the proportion of 39 to 1 be oftner repeated, the number of Tickets requifite for a Prize will ftill increafe with that repetition: Yet let the proportion of 39 to 1 be repeated never fo many times, nay an infinite number of times, the number of Tickets requifite for a Prize will never exceed $\frac{7}{10}$ of 39, that is about 27 or 28.

Therefore if the proportion of the Blanks to the Prizes be often repeated, as it ufually is in Lotteries; the number of Tickets requifite for one Prize or more, will always be found by taking $\frac{7}{10}$ of the proportion of the Blanks to the Prizes.

LEM-

The DOCTRINE *of* CHANCES.   37.

LEMMA.

TO *find how many Chances there are upon any number of Dice, each of them of the same given number of Faces, to throw any given number of Points.*

SOLUTION.

LET $p+1$ be the number of Points given, n the number of Dice, f the number of Faces in each Die: Make $p-f=q$; $q-f=r$; $r-f=s$; $s-f=t$ &c. and the number of Chances will be

$$+ \frac{p}{1} \times \frac{p-1}{2} \times \frac{p-2}{3} \ \&c.$$

$$- \frac{q}{1} \times \frac{q-1}{2} \times \frac{q-2}{3} \ \&c. \times \frac{n}{1}$$

$$+ \frac{r}{1} \times \frac{r-1}{2} \times \frac{r-2}{3} \ \&c. \times \frac{n}{1} \times \frac{n-1}{2}$$

$$- \frac{s}{1} \times \frac{s-1}{2} \times \frac{s-2}{3} \ \&c. \times \frac{n}{1} \times \frac{n-1}{2} \times \frac{n-2}{3}$$

$$+ \&c.$$

which Series ought to be continued till some of the Factors in each Product become either $= 0$, or Negative.

N. B. So many Factors are to be taken in each of the Products $\frac{p}{1} \times \frac{p-1}{2} \times \frac{p-2}{3}$ &c. $\frac{q}{1} \times \frac{q-1}{2} \times \frac{q-2}{3}$ &c. as there are Units in $n-1$.

Thus, for Example, let it be required to find how many Chances there are for throwing Sixteen Points with Four Dice.

$$+ \frac{15}{1} \times \frac{14}{2} \times \frac{13}{3} \qquad\qquad = + 455$$
$$- \frac{9}{1} \times \frac{8}{2} \times \frac{7}{3} \times \frac{4}{1} \qquad\quad = - 336$$
$$+ \frac{3}{1} \times \frac{2}{2} \times \frac{1}{3} \times \frac{4}{1} \times \frac{3}{2} \quad = + \ \ 6$$

But $455 - 336 + 6 = 125$; therefore One Hundred and Twenty Five is the number of Chances required.

F

Again,

Again, let it be required to find the number of Chances for throwing seven and Twenty Points with Six Dice.

$$+ \frac{26}{1} \times \frac{25}{2} \times \frac{24}{3} \times \frac{23}{4} \times \frac{22}{5} \qquad = + 65780$$

$$- \frac{19}{1} \times \frac{18}{2} \times \frac{17}{3} \times \frac{16}{4} \times \frac{15}{5} \times \frac{6}{1} \qquad = - 93204$$

$$+ \frac{11}{1} \times \frac{12}{2} \times \frac{11}{3} \times \frac{10}{4} \times \frac{9}{5} \times \frac{6}{1} \times \frac{5}{2} \qquad = + 30030$$

$$- \frac{8}{1} \times \frac{7}{2} \times \frac{6}{3} \times \frac{5}{4} \times \frac{4}{5} \times \frac{6}{1} \times \frac{5}{2} \times \frac{4}{3} = - 1120$$

Therefore $65780 - 93204 + 30030 - 1120 = 1666$ is the number required.

Let it be required to find the number of Chances for throwing Fifteen Points with Six Dice.

$$+ \frac{14}{1} \times \frac{13}{2} \times \frac{12}{3} \times \frac{11}{4} \times \frac{10}{5} \qquad = + 2002$$

$$- \frac{8}{1} \times \frac{7}{2} \times \frac{6}{3} \times \frac{5}{4} \times \frac{4}{5} \times \frac{6}{1} \qquad = - 336$$

But $2002 - 336 = 1666$, which is the number required.

Corol. All the Points equally distant from the extreams, that is, from the least and greatest number of Points that are upon the Dice, have the same number of Chances by which they may be produced; wherefore if the number of Points given be nearer to the greater Extream than to the less, let the number of Points given be subtracted from the sum of the Extreams, and work with the remainder, and the Operation will be shorten'd.

Thus, if it be required to find the number of Chances for throwing 27 Points with 6 Dice: Let 27 be subtracted from 42 the sum of the Extreams 6 and 36, and the Remainder being 15, it may be concluded that the number of Chances for throwing 27 Points is the same as for throwing 15.

Let it now be required to find in how many throws of 6 Dice one may undertake to throw 15 Points.

The number of Chances for throwing 15 Points being 1666; and the number of Chances for Failing being 44990; divide 44990 by 1666, the Quotient will be 27; Multiply 27
by

by 0.7, and the Product 18.9 will shew that the number of throws requisite to that effect is very near 19.

PROBLEM VI.

TO find how many Trials are necessary to make it Probable, that an Event will Happen twice, supposing that a *is the number of Chances for its Happening at any one Trial, and* b *the number of Chances for its Failing.*

SOLUTION.

LET x be the number of Trials: Therefore by what has been already Demonstrated, it will appear that $\overline{a+b}^x = 2b^x + 2axb^{x-1}$; or making $a, b :: 1, q$; $\overline{1+\frac{1}{q}}^x = 2 + \frac{2x}{q}$. Now if q be supposed $= 1$, x will be found $= 3$; and if q be supposed infinite, and also $\frac{x}{q} = z$, we shall have $z = \text{Log}: 2 + \text{Log}: \overline{1+z}$; in which Equation the value of z will be found $= 1.678$ very nearly. Therefore the value of x will always be between the Limits $3q$ and $1.678q$. But x will soon converge to the last of these Limits; therefore if x be not very small, it may in all cases be supposed $= 1.678q$. Yet if there be any Suspicion that the Value of x thus taken is too little, substitute this Value in the Original Equation $\overline{1+\frac{1}{q}}^x = 2 + \frac{2x}{q}$, and note that Errour. If it be worth taking notice of, then increase a little the value of x, and substitute again this new value in the room of x in the aforesaid Equation; and noting the new Errour, the value of x may be sufficiently corrected by applying the Rule which the Arithmeticians call double False Position.

EXAMPLE I.

TO find in how many throws of Three Dice one may undertake to throw Three Aces twice.

The number of all the Chances upon Three Dice being 216, out of which there is but one Chance for Three Aces, and 215 against it; Multiply 215 by 1.678, and the Product 360.7 will shew that the number of throws requisite to that effect will be 360 or very near it.

EX.

EXAMPLE II.

TO find in how many throws of Six Dice one may undertake to throw Fifteen Points twice.

The number of Chances for throwing Fifteen Points is 1666, the number of Chances for Miffing 44990; let 44990 be divided by 1666, the Quotient will be 27 very near: Wherefore the proportion of Chances for Throwing and Miffing Fifteen Points are as 1 to 27 refpectively; Multiply 27 by 1.678, and the Product 45.3 will fhew that the number of throws requifite to that effect will be 45 nearly.

EXAMPLE III.

IN a Lottery whereof the number of Blanks is to the number of Prizes as 39 to 1: To find how many Tickets muft be taken, to make it as Probable that two or more Benefits will be taken as not.

Multiply 39 by 1.678, and the Product 65.4 will fhew that no lefs than 65 Tickets will be requifite to that effect; tho' one might undertake upon an Equality of Chance to have one at leaft in 28.

PROBLEM VII.

TO *find how many Trials are neceffary to make it Probable that an Event will Happen Three, Four, Five, &c. times; fuppofing that* a *is the number of Chances for its Happening in any one Trial, and* b *the number of Chances for its Failing.*

SOLUTION.

LET x be the number of Trials requifite, then fuppofing, as before $a, b :: 1, q$, we fhall have the Equation $\overline{1 + \frac{1}{q}}^{x} = 2 \times \overline{1 + \frac{x}{q} + \frac{x}{1} \times \frac{x-1}{2qq}}$, in the cafe of the triple Event; or $\overline{1 + \frac{1}{q}}^{x} = 2 \times 1 + \frac{x}{q} + \frac{x}{1} \times \frac{x-1}{2qq} + \frac{x}{1} \times \frac{x-1}{2} \times \frac{x-2}{3qq}$, in the cafe of the quadruple Event: And the Law of the continuation of thefe Equations is manifeft. Now in the firft Equation if q be fuppofed $= 1$, then will x be $= 5$. If q be

fupsupposed infinite or pretty large in respect to Unity; then the aforesaid Equation, making $\frac{z}{q} = z$, will be changed into this, $z = $ Log. 2 $+$ Log. $\overline{1 + z + \frac{1}{2}zz}$; wherein z will be found nearly $= 2.675$. Wherefore x will always be between $5q$ and $2.675 q$.

Likewise in the second Equation, if q be supposed $= 1$, then will x be $= 7q$; but if x be supposed infinite, or pretty large in respect to Unity, then $z = $ Log. 2 $+$ Log. $\overline{1 + z + \frac{1}{2}zz + \frac{1}{6}z^3}$; whence z will be found nearly $= 3.6719$; Wherefore x will be between $7q$ and $3.6719q$.

If these Equations were continued, it would be found that the Limits of z converge continually to the proportion of two to one.

A TABLE of the Limits.

The Value of x will always be

For a single Event, between $1q$ and $0.693q$.
For a double Event, between $3q$ and $1.678q$.
For a triple Event, between $5q$ and $2.675q$.
For a quadruple Event, between $7q$ and $3.672q$.
For a quintuple Event, between $9q$ and $4.670q$.
For a sextuple Event, between $11q$ and $5.668q$.

If the number of Events contended for, as well as the number q be pretty large in respect to Unity; the number of Trials requisite for those Events to Happen n times, will be $\frac{2n-1}{2}q$ or barely nq.

PROBLEM VIII.

THree Gamesters A, B, C, *play together on this condition, that he shall win the Set who has soonest got a certain number of Games; the proportion of the Chances which each of them has to get any one Game assigned, or, which is the same thing, the proportion of their Skill, being respectively as* a, b, c. *Now after they have played some time, they find themselves in this circumstance, that* A *wants One Game of being up,* B *Two Games, and* C *Three; the whole Stake between them being supposed* 1 : *What is the Expectation of each?*

G   SOLU-

SOLUTION.

IN the Circumftance the Gamefters are in, the Set will be ended in Four Games at moft; let therefore $a + b + c$ be raifed
to the fourth Power, and it will be $a^4 + 4a^3b + 6aabb + 4ab^3$
$+ b^4 + 4a^3c + 12aabc + 4b^3c + 6aacc + 12abcc + 6bbcc$
$+ 4ac^3 + 12acbb + 4bc^3 + c^4$.

The Terms $a^4 + 4a^3b + 4a^3c + 6aacc + 12aabc + 12abcc$,
wherein the Dimenfions of a are equal to or greater than the
number of Games which A wants, wherein alfo the Dimenfions of b and c are lefs than the number of Games which
B and C want refpectively, are intirely Favourable to A, and
are part of the Numerator of his Expectation.

In the fame manner the Terms $b^4 + 4b^3c + 6bbcc$ are intirely Favourable to B.

And likewife the Terms $4bc^3 + c^4$ are intirely Favourable
to C.

The reft of the Terms are common, as Favouring partly
one of the Gamefters, partly one or both of the other: Wherefore thefe Terms are fo to be divided into their parts, that the
parts Favouring each Gamefter may be added to his Expectation.

Take therefore all the Terms which are common, *viz.*
$6aabb + 4ab^3 + 12abcc + 4ac^3$, and divide them actually
into their parts; that is 1°. $6aabb$ into *aabb, abab, abba, baab,
baba, bbaa.* Out of thefe Six parts, one part only, *viz. bbaa* will
be found to Favour B, for 'tis only in this Term that two Dimenfions of b are placed before one fingle Dimenfion of a, and
therefore the other Five parts belong to A; let therefore $5aabb$
be added to the Expectation of A, and $1aabb$ to the Expectation of B. 2° Divide $4ab^3$ into its parts, *abbb, babb, bbab, bbba.*
Of thefe parts there are two belonging to A, and the other
two to B; let therefore $2ab^3$ be added to the Expectation of
each. 3° Divide $12abbc$ into its parts; and eight of them will
be found Favourable to A, and four to B; let therefore $8abbc$
be added to the Expectation of A, and $4abbc$ to the Expectation of B. 4° Divide $4ac^3$ into its parts, three of which will
be found Favourable to A, and one to C; Add therefore $3ac^3$
to the Expectation of A, and $1ac^3$ to the Expectation of C.
Hence the Numerators of the feveral Expectations of A, B, C,
will be refpectively.   1.

1. $a^4 + 4a^3b + 4a^3c + 6aacc + 12aabc + 12abcc + 5aabb + 2ab^3 + 8abbc + 3ac^3$.

2. $b^4 + 4b^3c + 6bbcc + 12aabb + 2ab^3 + 4abcc$.

3. $4bc^3 + 1c^4 + 1ac^3$.

The common Denominator of all their Expectations being $\overline{a+b+c}|^4$.

Therefore if a, b, c are in a proportion of equality, the Odds of winning will be respectively as 57, 18, 6.

If n be the number of all the Games that are wanting, p the number of the Gamesters, a, b, c, d, &c. the proportion of the Chances which each Gamester has respectively to win any one Game assigned; let $a + b + c + d$ &c. be raised to the Power $n + 1 - p$, then proceed as before.

PROBLEM IX.

TWO Gamesters, A and B, *each having* 12 *Counters, play with three Dice, on condition, that if* 11 *Points come up,* B *shall give one Counter to* A; *if* 14, A *shall give one Counter to* B; *and that he shall be the winner who shall soonest get all the Counters of his adversary: What are the Probabilities that each of them has of winning?*

SOLUTION.

LET the number of Counters which each of them have be $= p$; and let a and b be the number of Chances they have respectively for getting a Counter each cast of the Dice: I say that the Probabilities of winning are respectively as a^p to b^p; or because in this case $p = 12$, $a = 27$, $b = 15$ as 27^{12} to 15^{12}, or as 9^{12} to 5^{12}, or as 282429536481 to 244140625, which is the proportion assigned by *M. Huygens*, but without any Demonstration:

Or more generally.

Let p be the number of the Counters of A, and q the number of the Counters of B; and let the proportion of the Chances be as a to b. I say that the proportions of the Probabilities which A has to get all the Counters of his adversary will be as $a^q \times \overline{a^p - b^p}$ to $b^p \times \overline{a^q - b^q}$.

DEMON-

DEMONSTRATION.

LET it be fuppofed that *A* has the Counters *E, F, G, H* &c. whofe number is *p*, and that *B* has the Counters *I, K, L* &c. whofe number is *q*: Moreover let it be fuppofed that the Counters are the thing play'd for, and that the Value of each of them is to the Value of the following as *a* to *b*, in fuch a manner that the laft Counter of *A* to the firft Counter of *B*, be ftill in that proportion. This being fuppofed, *A* and *B*, in every circumftance of their Play, may lay down two fuch Counters as may be proportional to the number of Chances each has to get a fingle Counter; for in the beginning of the Play *A* may lay down the Counter *H* which is the loweft of his Counters, and *B* the Counter *I* which is his higheft; but *H, I* :: *a, b*, therefore *A* and *B* play upon equal Terms. If *A* win of *B*, then *A* may lay down the Counter *I* which he has juft got of his adverfary, and *B* the Counter *K*; but *I, K*:: *a, b*, therefore *A* and *B* ftill play upon equal Terms. But if *A* lofe the firft time, then *A* may lay down the Counter *G*, and *B* the Counter *H*, which he but now got of his adverfary; but *G, H*:: *a, b*, and therefore they ftill Play upon equal Terms as before. So that as long as they Play together, they Play without advantage, or difadvantage, and confequently the Probabilities of winning are reciprocal to the Sums which they expect to win, that is, are proportional to the Sums they refpectively have before the Play begins. Whence the Probability which *A* has of winning all the Counters of *B*, is to the Probability which *B* has of winning all the Counters of *A*, as the Sum of the Terms *E, F, G, H* whofe number is *p*, to the Sum of the Terms *I, K, L* whofe number is *q*; that is, as $a^q \times \overline{a^p - b^p}$ to $b^p \times \overline{a^q - b^q}$: As will eafily appear if thofe Terms which are in Geometric Progreffion are actually fummed up by the known methods. Now the Probabilities of winning are not influenced by the fuppofition here made, of each Counter being to the following in the proportion of *a* to *b*; and therefore when thofe Counters are fuppofed of equal Value, or rather of no Value, but ferve only to mark the number of Stakes won or loft on either fide, the Probabilities of winning will be the fame as we have affigned.   R E-

REMARK I.

IF p and q, or either of them are large numbers, 'twill be convenient to work by Logarithms.

Thus, If A and B play a Guinea a Stake, and the number of Chances which A has to win each single Stake be 43, but the number of Chances which B has to win it be 40; and they oblige themselves to play till such time as 100 Stakes are won and lost.

$$
\begin{aligned}
\text{From the Logarithm of } 43 &= 1.6334685 \\
\text{Subtract the Logarithm of } 40 &= 1.6020600 \\
\hline
\text{Difference} &= 0.0314085
\end{aligned}
$$

Multiply this Difference by the number of Stakes to be play'd off, *viz.* 100; the Product will be 3.1048500, to which answers, in the Tables of Logarithms, the number 1383; wherefore the Odds that A shall win before B are 1383 to 1.

Now in all circumstances wherein A and B venture an equal Sum; the sum of the numbers expressing the Odds, is to their difference, as the Money play'd for, is to the Gain of the one, and the Loss of the other.

Therefore Multiplying 1382, difference of the numbers expressing the Odds, by 100, which is the sum ventured by each Man, and dividing the product by 1384 sum of the numbers expressing the Odds; the Quotient will be 99 Guineas, and about $18^{\text{sh.}} - 4\frac{1}{2}^{d}$, which consequently is to be estimated as the Gain of A.

REMARK II.

IF the number of Stakes which are to be won and lost be unequal, but the number of Chances to win and lose be equal; the Probabilities of winning will be reciprocally proportional to the number of Stakes to be won.

Thus, If A ventures Ten Stakes to win One; the Odds that he wins One before he loses Ten will be as 10 to 1.

But ten Chances to win One, and One Chance to lose ten, makes the Play perfectly equal.

H   There-

Therefore he that ventures many Stakes to win but few, has by it neither advantage nor difadvantage.

PROBLEM X.

TWO Gamefters, A *and* B *lay by* 24 *Counters, and play with Three Dice, on this condition* ; *that if* 11 *Points come up,* A *fhall take one Counter out of the heap* ; *if* 14, B *fhall take out one, and he fhall be reputed to win, who fhall foonefl get* 12 *Counters. What are the Probabilities of their winning* ?

This Problem differs from the preceding in this, that the play will be at an end in 23 Cafts of the Dice at moft, (that is of thofe Cafts which are favourable either to *A* or *B* :) Whereas in the preceding cafe, the Counters paffing continually from one Hand to the other, it will often Happen that *A* and *B* will be in fome of the fame circumftances they were in before, which will make the length of the play unlimited.

SOLUTION.

TAking *a* and *b* in the proportion of the Chances that there are to throw 11 and 14, let *a* + *b* be raifed to the 23*d* Power, that is to fuch Power as is denoted by the number of all the Counters wanting one : Then fhall the 12 firft Terms of that Power be to the 12 laft in the fame proportion as are the refpective Probabilities of winning.

PROBLEM XI.

THree Perfons A, B, C *out of a heap of* 12 *Counters, whereof four are White and Eight Black, draw blindfold one Counter at a time in this manner* ; A *begins to draw* ; B *follows* A ; C *follows* B ; *then* A *begins again* ; *and they continue to draw in the fame order, till one of them, who is to be reputed to win, draws the firft White one. What are the Probabilities of their winning* ?

SOLU.

SOLUTION.

LET n be the number of all the Counters, a the number of White ones, b the number of Black ones, and 1 the whole Stake or the sum play'd for.

1° Since A has a Chances for a White Counter, and b Chances for a Black one, it follows that the Probability of his winning is $\frac{a}{a+b}$ or $\frac{a}{n}$; Therefore the Expectation he has upon the Stake 1 arising from the circumstance he is in when he begins to draw is $\frac{a}{n} \times 1 = \frac{a}{n}$. Let it therefore be agreed amongst the adventurers that A shall have no Chance for a White Counter, but that he shall be reputed to have had a Black one, which shall actually be taken out of the heap, and that he shall have the sum $\frac{a}{n}$ paid him out of the Stake for an Equivalent. Now $\frac{a}{n}$ being taken out of the Stake, there will remain $1 - \frac{a}{n} = \frac{n-a}{n} = \frac{b}{n}$.

2° Since B has a Chances for a White Counter, and that the number of remaining Counters is $n-1$, his Probability of winning will be $\frac{a}{n-1}$. Whence his Expectation upon the remaining Stake $\frac{b}{n}$, arising from the circumstance he is now in, will be $\frac{ab}{n \times n-1}$. Suppose it therefore agreed that B shall have the sum $\frac{ab}{n \times n-1}$ paid him out of the Stake, and that a Black Counter be likewise taken out of the heap. This being done, the remaining Stake will be $\frac{b}{n} - \frac{ab}{n \times n-1}$, or $\frac{nb - b - ab}{n \times n-1}$; but $nb - ab = bb$; Wherefore the remaining Stake is $\frac{b \times b-1}{n \times n-1}$.

3° Since C has a Chances for a White Counter, and that the number of remaining Counters is $n-2$, his Probability of winning will be $\frac{a}{n-2}$: And therefore his Expectation upon the remaining Stake, arising from the circumstance he is now in, will be $\frac{b \times b-1 \times a}{n \times n-1 \times n-2}$ which we will likewise suppose to be paid him out of the Stake.

4° A may have out of the remainder $\frac{b \times b-1 \times b-2 \times a}{n \times n-1 \times n-2 \times n-3}$; and so of the rest till the whole Stake be exhausted.

There-

Therefore having written the following general Series, *viz.*

$$\tfrac{a}{s} + \tfrac{b}{s-1}P + \tfrac{b-1}{s-2}Q + \tfrac{b-2}{s-3}R + \tfrac{b-3}{s-4}S \text{ &c.}$$ wherein P, Q, R, S &c. denote the preceding Terms, take as many Terms of this Series as there are Units in $b + 1$, (for since b reprefents the number of Black Counters, the number of drawings cannot exceed $b + 1$) then take for *A* the firſt, fourth, feventh &c. Terms; for *B* take the ſecond, fifth, eighth &c. Terms; for *C* the third, fixth &c. and the fums of thofe Terms will be the refpective Expectations of *A, B, C*; or becaufe the Stake is fix'd, thefe fums will be proportional to their refpective Probabilities of winning.

Now to apply this to the prefent cafe, make $s = 12$, $a = 4$, $b = 8$, and the general Series will become

$$\tfrac{4}{12} + \tfrac{8}{11}P + \tfrac{7}{10}Q + \tfrac{6}{9}R + \tfrac{5}{8}S + \tfrac{4}{7}T + \tfrac{3}{6}V + \tfrac{2}{5}X + \tfrac{1}{4}Y:$$ Or multiplying the whole by 495, to take away the Fractions, the Series will be

$$165 + 120 + 84 + 56 + 35 + 20 + 10 + 4 + 1.$$

Therefore affign to *A* $165 + 56 + 10 = 231$; to *B* $120 + 35 + 4 = 159$; to *C* $84 + 20 + 1 = 105$, and their Probabilities of winning will be as 231, 159, 105, or as 77, 53, 35.

If there be never fo many Gamefters *A, B, C, D* &c. whether they take every one of them one Counter or more; or whether the fame or a different number of Counters; the Probabilities of winning may be determined by the fame general Series.

REMARK I.

THE preceding Series may in any particular cafe be fhorten'd; for if a is $= 1$, then the Series will be

$$\tfrac{1}{s} \times \overline{1 + 1 + 1 + 1 + 1 + 1 + 1} \text{ &c.}$$

Hence it may be obferved, that if the whole number of Counters be exactly divifible by the number of perfons concerned in the Play, and that there be but one fingle White Counter in the whole, there will be no advantage or difadvantage to any one of them from the fituation he is in, in refpect to the order of drawing.

If

If $a = 2$. then the Series will be

$$\frac{2}{n \times n - 1} \times n - 1 + n - 2 + n - 3 + n - 4 + n - 5 \ \&c.$$

If $a = 3$. then the Series will be

$$\frac{3}{n \times n - 1 \times n - 2} \times n - 1 \times n - 2 + n - 2 \times n - 3 + n - 3 \times n - 4 \ \&c.$$

If $a = 4$. then the Series will be

$$\frac{4}{n \times n - 1 \times n - 2 \times n - 3} \times n - 1 \times n - 2 \times n - 3 + n - 2 \times n - 3 \times n - 4 \ \&c.$$

Wherefore rejecting the common Multiplicators; the several Terms of these Series taken in due order will be Proportional to the several Expectations of any number of Gamesters. Thus in the case of this Problem where n is $= 12$, and $a = b$; the Terms of the Series will be

For *A.*		For *B.*		For *C.*	
$11 \times 10 \times 9 =$	990	$10 \times 9 \times 8 =$	720	$9 \times 8 \times 7 =$	504
$8 \times 7 \times 6 =$	336	$7 \times 6 \times 5 =$	210	$6 \times 5 \times 4 =$	120
$5 \times 4 \times 3 =$	60	$4 \times 3 \times 2 =$	24	$3 \times 2 \times 1 =$	6
	1386		954		630

Hence it follows, that the Probabilities of winning will be respectively as 1386, 954, 630; or dividing all by 18, as 77, 53, 35, as had been before determined.

REMARK II.

BUT if the Terms of the Series are many, it will be convenient to sum them up, by means of the following method, whose Demonstration may be had from the *Methodus Differentialis* of Sir *Isaac Newton*, printed in his *Analysis*.

Subtract every Term, but the first, from every following Term, and let the remainders be called *first Differences*; subtract in like manner every first difference from the following, and let the remainders be called *second Differences*; subtract again every one of these second differences from that which follows, and call the remainders *third Differences*; and so on, till the last differences become equal. Let the first Term be called a, the second b; the first of the first differences d', the

I   first

fi:ft of the second differences d'', the firft of the third diffe-rences d''' &c. and let the number of Terms which follow the firft be x, then will the fum of all thofe Terms be

$$a + xb + \frac{x}{1} \times \frac{x-1}{2} d' + \frac{x}{1} \times \frac{x-1}{2} \times \frac{x-2}{3} d'' +$$
$$\frac{x}{1} \times \frac{x-1}{2} \times \frac{x-2}{3} \times \frac{x-3}{4} d''' \ \&c.$$

N. B. If the numbers whofe fums are to be taken are the Products of two numbers, the fecond differences will be equal; if they are the Products of three, the third differences will be equal, and fo on. Therefore the number of Terms, which are to be taken after the firft, is to exceed only by Unity the number of Factors that enter the compofition of every Term.

It may alfo be obferved, that if thofe numbers are decrea-fing, it will be convenient to invert their order, and make that the firft which was the laft.

Thus, fuppofing the number of all the Counters to be 100, and the number of White ones 4 : Then the number of all the Terms belonging to *A, B, C* will be 97, the laft of which $3 \times 2 \times 1$ will belong to *A*, fince 97 being divided by 3, the remainder is 1. Therefore beginning from the loweft Term $3 \times 2 \times 1$, and taking every third Term, as alfo the differen-ces of thofe Terms, we fhall have the following Scheme

$$3 \times 2 \times 1 = \quad 6$$
$$6 \times 5 \times 4 = 120$$
$$\qquad\qquad\qquad 384$$
$$9 \times 8 \times 7 = 504 \qquad\quad 432$$
$$\qquad\qquad\qquad 816 \qquad\qquad 162$$
$$12 \times 11 \times 10 = 1320 \qquad 594$$
$$\qquad\qquad\qquad 1410$$
$$15 \times 14 \times 13 = 2730$$
&c.

From whence the Values of a, b, d', d'', d''', in the ge-neral Theorem, will be found to be refpectively 6, 120, 384, 432, 162; and confequently the fum of all thofe Terms will be

$6 +$

$$6 + x \times 120 + \frac{x}{1} \times \frac{x-1}{2} \times 384 + \frac{x}{1} \times \frac{x-1}{2} \times \frac{x-2}{3} \times 432$$
$$+ \frac{x}{1} \times \frac{x-2}{3} \times \frac{x-3}{4} \times 162, \quad \text{or}$$
$$6 + 31\tfrac{1}{2}x + 50\tfrac{1}{4}xx + 31\tfrac{1}{2}x^3 + 6\tfrac{3}{4}x^4, \quad \text{or}$$
$$\tfrac{3}{4} \times \overline{x+1} \times \overline{x+2} \times \overline{3x+1} \times \overline{3x+4}.$$

In like manner it will be found, that the sum of all the Terms which belong to B, the last of which is $5 \times 4 \times 3$ is

$$\tfrac{3}{4} \times \overline{x+1} \times \overline{x+2} \times \overline{3x+5} \times \overline{3x+8}.$$

And also that the sum of all the Terms belonging to C, the last of which is $4 \times 3 \times 2$, is

$$\tfrac{3}{4} \times \overline{x+1} \times \overline{x+2} \times \overline{9xx+27x+16}.$$

Now x in each case represents the number of Terms wanting one, which belong severally to A, B, C; wherefore making $x+1 = p$, their several Expectations will be respectively proportional to

$$p \times \overline{p+1} \times \overline{3p-2} \times \overline{3p+1}$$
$$p \times \overline{p+1} \times \overline{3p+2} \times \overline{3p+5}$$
$$p \times \overline{p+1} \times \overline{9pp+9p-2}.$$

Again, the number of all the Terms which belong to them all being 97, and A being to take first, it follows, that p in the first case is $= 33$, in the other two $= 32$.

Therefore the several Expectations of A, B, C will be respectively proportional to 41225, 39592, 38008.

If the number of all the Counters were 500, and the number of the White ones still 4; then the number of all the Terms representing the Expectations of A, B, C would be 497. Now this number being divided by 3, the Quotient is 165, and the Remainder 2: From whence it follows, that the last Term $3 \times 2 \times 1$ will belong to B, the last but one $4 \times 3 \times 2$ to A, and the last but two to C; it follows also, that for B, and A, p must be interpreted by 166, but for C by 165.

The

The GAME of BASSETE.

RULES of the PLAY.

THE Dealer, otherwife called the *Banker*, holds a Pack of 52 Cards, and having fhuffled them, he turns the whole Pack at once, fo as to difcover the laft Card; after which he lays down by Couples all the Cards.

The Setter, otherwife called the *Ponte*, has 13 Cards in his hand, one of every fort, from the King to the Ace, which 13 Cards are called a *Book*; out of this Book he takes one Card or more at pleafure, upon which he lays a Stake.

The Ponte may at his choice, either lay down his Stake before the Pack is turned, or immediately after it is turned; or after any number of Couples are drawn.

The firft cafe being particular fhall be calculated by it felf; but the other two are comprehended under the fame Rules.

Suppofing the Ponte to lay down his Stake after the Pack is turned, I call 1, 2, 3, 4, 5 &c. the places of thofe Cards which follow the Card in view, either immediately after the Pack is turned, or after any number of Couples are drawn.

If the Card upon which the Ponte has laid a Stake comes out in any odd place, except the firft, he wins a Stake equal to his own.

If the Card upon which the Ponte has laid a Stake comes out in any even place except the fecond, he lofes his Stake.

If the Card of the Ponte comes out in the firft place, he neither wins nor lofes, but takes his own Stake again.

If the Card of the Ponte comes out in the fecond place, he does not lofe his whole Stake, but only a part of it, *viz,* a half; which to make the calculation more general we will call *y.* In this cafe the Ponte is faid to be *Faced.*

When the Ponte chufes to come in after any number of Couples are down; if his Card happens to be but once in the Pack, and is the very laft of all, there is an exception from the general Rule: for tho' it comes out in an odd place which fhould intitle him to win a Stake equal to his own, yet he neither wins nor lofes from that circumftance, but takes back his own Stake.

PRO-

PROBLEM XII.

TO *Eſtimate at* Baſſete *the loſs of the* Ponte *under any circumſtances of Cards remaining in the Stock, when he lays
his Stake, and of any number of times that his Card is repeated
in it.*

The Solution of this Problem containing Four Caſes, *viz.*
of the Ponts Card being once, twice, three or four times in
the Stock; we will give the Solution of all theſe Caſes ſeverally.

SOLUTION of the firſt Caſe.

THe *Ponte* has the following Chances to win or loſe, according to the place his Card is in.

1	1 Chance for winning	0
2	1 Chance for loſing	*y*
3	1 Chance for winning	1
4	1 Chance for loſing	1
5	1 Chance for winning	1
6	1 Chance for loſing	1
*	1 Chance for winning	0

It appears by this Scheme that he has as many Chances
to win 1 as to loſe 1, and that there are two Chances
for neither winning nor loſing, *viz.* the firſt and laſt, and
therefore that his only Loſs is upon account of his being
Faced: From which 'tis plain that the number of Cards covered by that which is in view being called *n*, his Loſs will
be $\frac{y}{n}$, or $\frac{1}{2n}$ ſuppoſing $y = \frac{1}{2}$.

SOLUTION of the ſecond Caſe.

By the firſt Remark belonging to the XI*th* Problem it appears
that the Chances which the Ponte has to win or loſe are
proportional to the numbers, *n*—1, *n*—2, *n*—3 &c. Therefore his Chances for winning and loſing may be expreſſed
by the following Scheme.

K

1	$n-1$ Chances for winning	o
2	$n-2$ Chances for losing	y
3	$n-3$ Chances for winning	1
4	$n-4$ Chances for losing	1
5	$n-5$ Chances for winning	1
6	$n-6$ Chances for losing	1
7	$n-7$ Chances for winning	1
8	$n-8$ Chances for losing	1
9	$n-9$ Chances for winning	1
*	1 Chance for losing	1

Now setting aside the first and second number of Chances, it will be found that the difference between the 3*d* and 4*th* is $= 1$, and that the difference between the 5*th* and 6*th* is $= 1$. The difference between the 7*th* and 8*th* also is $= 1$, and so on. But the number of differences is $\frac{n-3}{2}$, and the sum of all the Chances is $\frac{n}{1} \times \frac{n-1}{2}$. Wherefore the Gain of the Ponte is $\frac{n-3}{n \times n - 1}$; but his Loss upon account of the *Face* is $\overline{n-2} \times y$ divided by $\frac{n}{1} \times \frac{n-1}{2}$, or $\frac{2n-4 \times y}{n \times n - 1}$: Hence it may be concluded that his Loss upon the whole is $\frac{2n-4 \times y - \overline{n-3}}{n \times n - 1}$, or $\frac{1}{n \times n - 1}$ supposing $y = \frac{1}{2}$.

That the number of Differences is $\frac{n-3}{2}$ will be made evident from two considerations.

First, the Series $n-3$, $n-4$, $n-5$ &c. decreases in Arithmetic Progression, the difference of its Terms being Unity, and the last Term also Unity, therefore the number of its Terms is equal to the first Term $n-3$: But the number of Differences is one half of the number of Terms, therefore the number of Differences will be $\frac{n-3}{2}$.

Secondly, It appears by the XI*th* Problem, that the number of all the Terms including the two first is always $b+1$; But b in this case is $= 2$. Therefore the number of all the Terms is $n-1$, from which excluding the two first, the number of remaining Terms will be $n-3$, and consequently the number of Differences will be $\frac{n-3}{2}$.

That the sum of all the Terms is $\frac{n}{1} \times \frac{n-1}{2}$, is evident also from two different considerations.   First,

First, In any Arithmetic Progression whereof the first Term is $n-1$, the difference Unity, and the last Term also Unity, the sum of the Progression will be $\frac{n}{1} \times \frac{n-1}{2}$.

Secondly, the Series $\frac{2}{n \times n-1} \times \overline{n-1 + n-2 + n-3}$ &c. belonging to the preceding Problem, expresses the sum of the Probabilities of winning, which belong to the several Gamesters in the case of two White Counters, when the number of all the Counters is *n*. It therefore expresses likewise the sum of the Probabilities of winning, which belong to the Ponte or Banker in the present case: But this sum must always be equal to Unity, it being a certainty that the Ponte or Banker must win; supposing therefore that $n-1, n-2$, $n-3$ &c. is $= S$. we shall have the Equation $\frac{2S}{n \times n-1} = 1$. Therefore $S = \frac{n}{1} \times \frac{n-1}{2}$.

SOLUTION of the third Case.

By the first Remark of the XI*th* Problem it appears that the Chances which the Ponte has to win and lose, may be expressed by the following Scheme:

1	$n-1 \times n-2$ for winning	o
2	$n-2 \times n-3$ for losing	y
3	$n-3 \times n-4$ for winning	1
4	$n-4 \times n-5$ for losing	1
5	$n-5 \times n-6$ for winning	1
6	$n-6 \times n-7$ for losing	1.
7	$n-7 \times n-8$ for winning	1.
8	$n-8 \times n-9$ for losing	1
*	2×1 for winning	1

Setting aside the first, second and last number of Chances; it will be found that the difference between the 3*d* and 4*th* is $2n-8$; the difference between the 5*th* and 6*th* $2n-12$, the difference between the 7th and 8*th* $2n-16$. &c. Now these differences constitute an Arithmetic Progression, whereof the first Term is $2n-8$, the common difference 4, and the last Term 6, being the difference between 4×3 and 3×2. Wherefore the sum of this Progression is $\frac{n-3}{1} \times \frac{n-2}{2}$, to which adding the last Term 2×1, which is favourable to the Ponte.

Ponte, the sum total will be $\frac{n-3}{1} \times \frac{n-3}{2}$. But the sum of all the Chances is $\frac{n}{1} \times \frac{n-1}{1} \times \frac{n-3}{3}$; as may be concluded from the first Remark of the preceding Problem: Therefore the Gain of the Ponte is $\frac{3 \times n-1 \times n-1}{2 \times n \times n-1 \times n-2}$. But his Loss upon account of the Face is $\frac{6 \times n-2 \times n-3 \times y}{2 \times n \times n-1 \times n-2}$. Consequently his Loss upon the whole will be $\frac{6 \times n-3 \times n-3 \times y-3 \times n-3 \times n-3}{2 \times n \times n-1 \times n-2}$ or $\frac{3 n-9}{2 \times n \times n-1 \times n-2}$, Supposing $y = \frac{1}{2}$.

SOLUTION of the fourth Case.

The Chances of the Ponte may be expressed by the following Scheme.

1	$n-1 \times n-2 \times n-3$	for winning	0
2	$n-2 \times n-3 \times n-4$	for losing	y
3	$n-3 \times n-4 \times n-5$	for winning	1
4	$n-4 \times n-5 \times n-6$	for losing	1
5	$n-5 \times n-6 \times n-7$	for winning	1
6	$n-6 \times n-7 \times n-8$	for losing	1
7	$n-7 \times n-8 \times n-9$	for winning	1
*	$3 \times 2 \times 1$	for losing	1

Setting aside the first and second numbers of Chances, and taking the differences between the 3d and 4th, 5th and 6th; 7th and 8th, the last of these differences will be found to be 18. Now if the number of these differences be p, and we begin from the last 18, their sum, from the second Remark of the preceding Problem, will be collected to be $p \times \overline{p+1} \times 4p+5$: And the number p in this case being $\frac{n-3}{2}$, the sum of these differences will be $\frac{n-1}{2} \times \frac{n-3}{2} \times \frac{2n-5}{1}$. But the sum of all the Chances is $\frac{n}{1} \times \frac{n-1}{1} \times \frac{n-2}{1} \times \frac{n-3}{4}$; wherefore the Gain of the Ponte is $\frac{n-5 \times n-3 \times 2 n-5}{n \times n-1 \times n-2 \times n-3} \cdot 2$: now his Loss upon account of the Face is $\frac{n-1 \times n-3 \times n-4 \times 4 y}{n \times n-1 \times n-2 \times n-3}$ and therefore his Loss upon the whole is $\frac{4 \times n-2 \times n-4 \times y-n-5 \times 2 n-5}{n \times n-1 \times n-2}$ or $\frac{3 n-9}{n \times n-1 \times n-2}$, making $y = \frac{1}{2}$.

There

There still remains the single Case to be considered, *viz.* what the Loss of the Ponte is, when he lays a Stake before the Pack is turned up; but there will be no difficulty in it after what we have said, the difference between this Case and the rest being only that he may be Faced by the first Card discovered, which will make his Loss to be $\frac{3n-6}{nn-nn-3}$, that is, about $\frac{1}{866}$ part of his Stake.

Those who are desirous to try, by a kind of Mechanical Operation, the truth of the Rules which have been given for determining the Loss of the Ponte in any Case, may do it in the following manner. Suppose for Instance it were required to find the Loss of the Ponte when his Card is twice in the Stock, and there are five Cards remaining in the hands of the Banker beside the Card in View. Let them be disposed according to this Scheme.

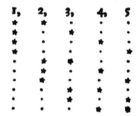

Where the places filled with Afterisks shew all the Various Positions which the Ponte's Card may obtain; it is evident that the Ponte has four Chances for neither winning nor losing, three Chances for the Face or for losing $\frac{1}{2}$, two Chances for winning 1, and one Chance for losing 1; and consequently that his Loss is $\frac{1}{2}$ to be distributed into 10 parts, the number of all the Chances being 10, which will make his Loss to be $\frac{1}{20}$. Likewise if the number of Cards that are covered by the first were seven, it would be found that the Ponte would have six Chances for neither winning nor losing, five Chances for the Face, four Chances for winning 1, three Chances for losing 1, two Chances again for winning 1, and one Chance for losing 1, which would make his Loss to be $\frac{1}{42}$. And the like may be done for any other Case whatsoever.

L

From what has been faid, a Table may eafily be compofed, fhewing the feveral Loffes of the Ponte in whatever circumftance he may happen to be.

A TABLE for *BASSETE.*

N	1	2	3	4
52	* * *	* * *	* * *	866
51	* * *	* * *	1735	867
49	98	2352	1602	801
47	94	2162	1474	737
45	90	1980	1351	675
43	86	1806	1234	617
41	82	1640	1122	561
39	78	1482	1015	507
37	74	1332	914	457
35	70	1190	818	409
33	66	1056	727	363
31	62	930	642	321
29	58	812	562	281
27	54	702	487	243
25	50	600	418	209
23	46	506	354	177
21	42	420	295	147
19	38	342	242	121
17	34	272	194	97
15	30	210	151	75
13	26	156	114	57
11	22	110	82	41
9	18	72	56	28
7	14	42	25	17

The Ufe of this Table will be beft explained by one or
two Examples.   Exam-

EXAMPLE I.

LET it be propofed to find the Lofs of the Ponte when there are 26 Cards remaining in the Stock, and his Card is twice in it.

In the Column *N* find the number 25, which is lefs by one than the number of Cards remaining in the Stock: Over againft it, and under the number 2, which is at the head of the fecond Column, you will find 600; which is the Denominator of a Fraction whofe Numerator is Unity, and which fhews that his Lofs in that circumftance is one part in fix hundred of his Stake.

EXAMPLE II.

TO find the Lofs of the Ponte when there is eight Cards remaining in the Stock, and his Card is three times in it.

In the Column *N* find the number 7, lefs by one than the number of Cards remaining in the Stock : Over againft 7, and under the number 3 in the third Column, you will find 35; which denotes that his Lofs is one part in thirty five of his Stake.

Corollary I. 'Tis plain from the conftruction of the Table, that the fewer Cards are in the Stock, the greater is the Lofs of the Ponte.

Corollary II. The leaft Lofs of the Ponte, under the fame circumftances of Cards remaining in the Stock, is when his Card is but twice in it; the next greater when three times; ftill greater when four times, but his greateft Lofs when 'tis but once.

If the Lofs upon the Face were varied, 'tis plain that in all the like circumftances, the Lofs of the Ponte would vary accordingly, but it would be eafie to compofe other Tables to anfwer that Variation, fince the quantity y, which has been affumed to reprefent that Lofs may be interpreted at pleafure. For inftance, when the Lofs upon the Face is $\frac{1}{2}$, it has been found in the Cafe of 7 Cards covered remaining in the Stock, and the Card of the Ponte being twice in it, that his Lofs would be $\frac{1}{42}$, but upon fuppofition of its being $\frac{2}{3}$, it will be found to be $\frac{4}{63}$.

The

The GAME of PHARAON.

THE Calculation for *Pharaon* is much like the preceding, the reasonings about it being the same; therefore I think it will be sufficient to lay down the Rules of the Play, and the Scheme of the Calculation.

RULES of the PLAY.

First, The Banker holds a Pack of 52 Cards.

Secondly, He draws the Cards one after the other, laying them alternately to his right and left hand.

Thirdly, the Ponte may at his choice set one or more Stakes upon one or more Cards either before the Banker has begun to draw the Cards, or after he has drawn any number of Couples, which are commonly called *Pulls.*

Fourthly, The Banker wins the Stake of the Ponte, when the Card of the Ponte comes out in an odd place on his right hand; but loses as much to the Ponte when it comes out in an even place on his left hand.

Fifthly, The Banker wins half the Ponte's Stake, when in the same Pull the Card of the Ponte comes out twice.

Sixthly, When the Card of the Ponte, being but once in the Stock, happens to be the last, the Ponte neither wins nor loses.

Seventhly, The Card of the Ponte being but twice in the Stock, and the two last Cards happening to be his Cards, he then loses his whole Stake.

PROBLEM XIII.

TO *Find at* Pharaon *the Gain of the Banker, in any Circumstance of Cards remaining in the Stock, and of the number of times that the Ponte's Card is contained in it.*

This Problem, containing four Cases, that is, when the Card of the Ponte is once, twice, three or four times in the Stock; we shall give the Solution of these four Cases severally.

SOLU-

SOLUTION of the firſt Caſe.

The Banker has the following number of Chances for winning and loſing, *viz.*

1	1 Chance for winning	1
2	1 Chance for loſing	1
3	1 Chance for winning	1
4	1 Chance for loſing	1
5	1 Chance for winning	1
*	1 Chance for loſing	0

Therefore the Gain of the Banker is $\frac{1}{n}$. Suppoſing *n* to be the number of Cards in the Stock.

SOLUTION of the ſecond Caſe.

The Banker has the following Chances for winning and loſing, *viz.*

1	$n-2$ Chances for winning	1
	1 Chance for winning	y
2	$n-2$ Chances for loſing	1
3	$n-4$ Chances for winning	1
	1 Chance for winning	y
4	$n-4$ Chances for loſing	1
5	$n-6$ Chances for winning	1
	1 Chance for winning	y
6	$n-6$ Chances for loſing	1
7	$n-8$ Chances for winning	1
	1 Chance for winning	y
8	$n-8$ Chances for loſing	1
*	1 Chance for winning	1

Therefore the Gain of the Banker is $\frac{n-2\times y+2}{n\times n-1}$, or $\frac{n+2}{n\times n-1}$, suppoſing $y=\frac{1}{2}$.

M

The

The only thing that deferves to be explained here, is this; how it comes to pafs that whereas at *Baffete* the firft number of Chances for winning was reprefented by $n-1$, here 'tis reprefented by $n-2$.   To anfwer this it muft be remember'd, that according to the Law of this Play, if the Ponte's Card comes out in an odd place, the Banker is not thereby entitled to the Ponte's whole Stake: For if it fo happens that his Card comes out again immediately after, the Banker wins but one half of it.   Therefore the number $n-1$ is divided into two parts $n-2$ and 1, whereof the firft is proportional to the Probability which the Banker has for winning the whole Stake of the Ponte; and the fecond is proportional to the Probability of his winning the half of it.

SOLUTION of the third Cafe.

The number of Chances which the Banker has for winning, and lofing are as follow;

1	{	$n-2 \times n-3$ Chances for winning	1
	{	$2 \times n-2$ Chances for winning	y
2		$n-2 \times n-3$ Chances for lofing	1
3	{	$n-4 \times n-5$ Chances for winning	1
	{	$2 \times n-4$ Chances for winning	y
4		$n-4 \times n-5$ Chances for lofing	1
5	{	$n-6 \times n-7$ Chances for winning	1
	{	$2 \times n-6$ Chances for winning	y
6		$n-6 \times n-7$ Chances for lofing	1
7	{	$n-8 \times n-9$ Chances for winning	1
	{	$2 \times n-8$ Chances for winning	y
*		2×1 Chances for lofing	1

Therefore the Gain of the Banker is $\frac{3y}{2 \times \overline{n-1}}$, or $\frac{1}{4 \times \overline{n-1}}$ fuppofing $y = \frac{1}{2}$.

The number of Chances for the Banker to win is divided into two parts, whereof the firft exprefles the Chances he has for winning the whole Stake of the Ponte, and the fecond for winning the half thereof.

Now

Now for determining exactly thefe two parts, it may be confidered, that in the firft Pull the number of Chances for the firft Card to be the Ponte's is $n-1 \times n-2$; alfo that the number of Chances for the fecond to be the Ponte's but not the firft, is $n-2 \times n-3$: Wherefore the number of Chances for the firft to be the Ponte's and not the fecond, is likewife $n-2 \times n-3$. Hence it follows, that if from the number of Chances for the firft Card to be the Ponte's, *viz.* from $n-1 \times n-2$ there be fubtracted the number of Chances for the firft to be the Ponte's and not the fecond, *viz.* $n-2 \times n-3$, there will remain the number of Chances for both firft and fecond Cards to be the Ponte's, *viz.* $2 \times n-2$ and fo for the reft.

SOLUTION of the fourth Cafe.

The number of Chances which the Banker has for winning and lofing, are as follows;

1	$\{$	$n-2 \times n-3 \times n-4$ Chances for winning	1
		$3 \times n-2 \times n-3$ Chances for winning	y
2		$n-2 \times n-3 \times n-4$ Chances for lofing	1
3	$\{$	$n-4 \times n-5 \times n-6$ Chances for winning	1
		$3 \times n-4 \times n-5$ Chances for winning	y
4		$n-4 \times n-5 \times n-6$ Chances for lofing	1
5	$\{$	$n-6 \times n-7 \times n-8$ Chances for winning	1
		$3 \times n-6 \times n-7$ Chances for winning	y
6		$n-6 \times n-7 \times n-8$ Chances for lofing	1
7	$\{$	$n-8 \times n-9 \times n-10$ Chances for winning	1
		$3 \times n-8 \times n-9$ Chances for winning	y
8		$n-8 \times n-9 \times n-10$ Chances for lofing	1
*	$\{$	$2 \times 1 \times 0$ Chances for winning	1
		$3 \times 2 \times 1$ Chances for winning	1
		$2 \times 1 \times 0$ Chances for lofing	1

Therefore the Gain of the Banker is $\frac{2n-5}{n-1 \cdot n-3} y$, or $\frac{2n-5}{2 \cdot n-1 \cdot n-3}$, fuppofing $y = \frac{1}{2}$.

A TA-

The DOCTRINE *of* CHANCES.

A TABLE for *PHARAON.*

N	1	2	3	4
52	* * *	* * *	* * *	50
50	* * *	94	65	48
48	48	90	62	46
46	46	86	60	44
44	44	82	57	42
42	42	78	54	40
40	40	74	52	38
38	38	70	49	36
36	36	66	46	34
34	34	62	44	32
32	32	58	41	30
30	30	54	38	28
28	28	50	36	26
26	26	46	33	24
24	24	42	30	22
22	22	38	28	20
20	20	34	25	18
18	18	30	22	16
16	16	26	20	14
14	14	22	17	12
12	12	18	14	10
10	10	14	12	8
8	8	11	9	6

The numbers of the foregoing Table, as well as those of the Table for *Baſſete*, are ſufficiently exact to give at firſt view an Idea of the advantage of the Banker in all circumſtances: But if an abſolute degree of exactneſs be required, it will be eaſily obtained from the Rules given at the end of each Caſe.

PRO-

PROBLEM XIV.

IF A, B, C throw in their turns a regular Ball, having four White Faces and eight Black ones; and he be to be reputed to win who shall first bring up one of the White Faces: It is demanded what is the proportion of their respective Probabilities of winning?

SOLUTION.

THe method of reasoning in this Problem is exactly the same with that which we made use of in the Solution of the XI*th* Problem : But whereas the different throws of the Ball do not diminish the number of its Faces; in the room of the Quantities $b-1$, $b-2$, $b-3$ &c. $n-1$, $n-2$, $n-3$ &c. employed in the Solution of the aforesaid Problem, we must substitute b and n respectively, and the Series belonging to that Problem will be changed into the following, *viz.*

$$\frac{a}{n} + \frac{ab}{nn} + \frac{abb}{n^3} + \frac{ab^3}{n^4} + \frac{ab^4}{n^5} + \frac{ab^5}{n^6} \text{ \&c.}$$

which is to be continued infinitely : Then taking every third Term thereof, the respective Expectations of *A*, *B*, *C* will be expressed by the three following Series.

$$\frac{a}{n} + \frac{ab^3}{n^4} + \frac{ab^6}{n^7} + \frac{ab^9}{n^{10}} + \frac{ab^{12}}{n^{13}} \text{ \&c.}$$

$$\frac{ab}{nn} + \frac{ab^4}{n^5} + \frac{ab^7}{n^8} + \frac{ab^{10}}{n^{11}} + \frac{ab^{13}}{n^{14}} \text{ \&c.}$$

$$\frac{abb}{n^3} + \frac{ab^5}{n^6} + \frac{ab^8}{n^9} + \frac{ab^{11}}{n^{12}} + \frac{ab^{14}}{n^{15}} \text{ \&c.}$$

But the Terms of which each Series is compounded are in Geometric Progression, and the Ratio of each Term to the following the same in each of them; Wherefore the Sums of these Series are in the same proportion as their first Terms, *viz.* as $\frac{a}{n}$, $\frac{ab}{nn}$, $\frac{abb}{n^3}$ or as nn, bn, bb; that is, in the present Case, as 144, 96, 64, or as 9, 6, 4. Hence the respective Probabilities of winning will be likewise as the numbers 9, 6, 4.

Corollary I. If there be any other number of Gamesters *A*, *B*, *C*, *D* &c. playing on the same conditions as above;

N   take

take as many Terms in the Ratio of *n* to *b* as there are Gamesters, and those Terms will respectively denote the several Expectations of each Gamester.

Corollary II. If there be any number of Gamesters *A, B, C, D* &c. playing on the same conditions as above; with this difference only, that all the Faces of the Ball are mark'd by particular Figures, 1, 2, 3, 4 &c. and that a certain number *p* of those Faces shall intitle *A* to be the winner; and that likewise any other number of them, as *q, r, s, t* &c. shall respectively intitle *B, C, D, E* &c. to be winners: Make $n-p=a$, $n-q=b, n-r=c, n-s=d, n-t=e$ &c. then in the following Series,

$$\frac{p}{n} + \frac{qa}{nn} + \frac{rab}{n^3} + \frac{sabc}{n^4} + \frac{tabcd}{n^5} \text{ &c.}$$

the Terms taken in due order shall represent the several Probabilities of winning.

For if the Law of the Play be such, that every Man having once play'd in his turn, shall begin again regularly in the same manner, and that continually till such time as one of them wins: Then take as many Terms of the Series as there are Gamesters, and those Terms shall represent the respective Probabilities of winning.

And if it were the Law of the Play, that every Man should play several times together, for instance twice: Then taking for *A* the two first Terms, for *B* the two following, and so on; each Couple of Terms shall represent their respective Probabilities of winning; observing that now *p* and *q* are equal, as also *r* and *s.*

But if the Law of the Play should be Irregular, then you must take for each Man as many Terms of the Series as will answer that Irregularity, and continue the Series till such time as it gives a sufficient Approximation.

Yet, if at any time the Law of the Play having been Irregular should afterwards recover its Regularity, the Probabilities of winning will (with the help of this Series) be determined by finite expressions.

Thus, if it should be the Law of the Play, that two Men *A* and *B*, having play'd irregularly for ten times together, should afterwards play alternately each in his turn: Distribute the ten first Terms of the Series between them, according

ing

ing to their order of playing; and having subtracted the sum of those Terms from Unity, divide the remainder of it between them, in the proportion of the two following Terms, which add respectively to the shares they had before: Then shall the two parts of Unity which *A* and *B* have thus obtained, be proportional to their respective Probabilities of winning.

Of Permutations and Combinations.

Permutations are the Changes which several things can receive in the different Orders in which they may be placed, being considered as taken two and two, three and three, four and four, &c.

Combinations are the various Conjunctions which several things may receive without any respect to Order, being taken two and two, three and three, four and four, &c.

LEMMA.

IF the Probability that an Event shall Happen be $\frac{1}{r}$, and if that Event being supposed to have Happened, the Probability of anothers Happening be $\frac{1}{s}$; the Probability of both Happening will be $\frac{1}{r} \times \frac{1}{s}$ or $\frac{1}{rs}$. This having been already Demonstrated in the Introduction, will not require any farther proof.

PROBLEM XV.

ANY number of Things a, b, c, d, e, f being given, out of which Two are taken as it happens: To find the Probability that any one of them, as a, shall be the first taken, and any other, as b, the second.

SOLUTION.

THE number of Things in this Example being Six, it follows that the Probability of taking *a* in the first place is $\frac{1}{6}$: Let *a* be considered as taken, then the Probability of taking *b* will be $\frac{1}{5}$; wherefore the Probability of taking first *a* and then *b* is $\frac{1}{6} \times \frac{1}{5} = \frac{1}{30}$

Corol.

Corollary. Since the taking of *a* in the firſt place and *b* in the ſecond, is but one ſingle Caſe of thoſe by which Six Things may change their Order, being taken two and two; it follows, that the number of Changes or Permutations of Six Things taken two and two muſt be 30.

Generally, Let *n* be any number of Things; the Probability of taking *a* in the firſt place and *b* in the ſecond, will be $\frac{1}{n \times n-1}$; and the number of Permutations of thoſe Things taken two by two will be $n \times n-1$.

PROBLEM XVI.

ANY *number* n *of Things* a, b, c, d, e, f *being given, out of which Three are taken as it Happens: To find the Probability that* a *ſhall be the firſt taken,* b *the ſecond and* c *the third.*

SOLUTION.

THe Probability of taking *a* in the firſt place is $\frac{1}{6}$: Let *a* be conſidered as taken; the Probability of taking *b* will be $\frac{1}{5}$: Suppoſe both *a* and *b* taken, the Probability of taking *c* will be $\frac{1}{4}$. Wherefore the Probability of taking firſt *a*, then *b*, and thirdly *c*, will be $\frac{1}{6} \times \frac{1}{5} \times \frac{1}{4} = \frac{1}{120}$.

Corollary. Since the taking of *a* in the firſt place, *b* in the ſecond, and *c* in the third, is but one ſingle Caſe of thoſe by which Six Things may change their Order, being taken three and three; it follows, that the number of Changes or Permutations of Six Things, taken three and three, muſt be $6 \times 5 \times 4 = 120$.

Generally, If *n* be any number of Things; the Probability of taking *a* in the firſt place, *b* in the ſecond and *c* in the third will be $\frac{1}{n} \times \frac{1}{n-1} \times \frac{1}{n-2}$. And the number of Permutations of three Things will be $n \times n-1 \times n-2$.

General COROLLARY.

The number of Permutations of *n* Things, out of which as many are taken together as there are Units in *p*, will be $n \times n-1 \times n-2 \times n-3$, &c. continued to ſo many Terms as there are Units in *p*.

Thus,

Thus, the number of Permutations of Six Things taken four and four, will be $6 \times 5 \times 4 \times 3 = 360$. Likewise the number of Permutations of Six Things taken all together will be $6 \times 5 \times 4 \times 3 \times 2 \times 1 = 720$.

PROBLEM XVII.

TO *Find the Probability that any number of Things, whereof some are repeated several times, shall all be taken in any Order proposed: For Instance, that* aabbbcccc *shall be taken in the Order wherein they are written.*

SOLUTION.

THe Probability of taking *a* in the first place is $\frac{2}{9}$: Supposing one *a* to be taken; the Probability of taking the other is $\frac{1}{8}$. Let now the two first Letters be supposed to be taken, the Probability of taking *b* will be $\frac{3}{7}$: Let this also be supposed taken, the Probability of taking another *b* will be $\frac{2}{6}$: Let this likewise be supposed taken, the Probability of taking the third *b* will be $\frac{1}{5}$; after which there remaining nothing but the Letter *c*, the Probability of taking it becomes a certainty, and consequently is equal to Unity. Wherefore the Probability of taking all those Letters in the Order given is $\frac{2}{9} \times \frac{1}{8} \times \frac{3}{7} \times \frac{2}{6} \times \frac{1}{5}$.

Corollary. Therefore the number of Permutations which the Letters aabbbcccc may receive, being taken all together will be $\frac{9 \times 8 \times 7 \times 6 \times 5}{2 \times 1 \times 3 \times 2 \times 1}$.

Generally. The number of Permutations which any number *n* of Things may receive, being taken all together, whereof the first sort is repeated *p* times, the second *q* times, the third *r* times, the fourth *s* times, &c. will be the Series $n \times n-1 \times n-2 \times n-3 \times n-4$, &c. continued to so many Terms as there are Units in $p+q+r$ or $n-s$, divided by the Product of the following Series, viz. $p \times p-1 \times p-2$, &c. $\times q \times q-1 \times q-2$, &c. $\times r \times r-1 \times r-2$, &c. whereof the first must be continued to so many Terms as there are Units in *p*; the second, to so many Terms as three are Units in *q*; the third, to so many Terms as there are Units in *r* &c.

O   PROB-

PROBLEM XVIII.

ANY *number of Things* a, b, c, d, e, f *being given: To find the Probability that, in taking two of them as it may Happen, both* a *and* b *shall be taken independently, or without any regard to Order.*

SOLUTION.

THE Probability of taking *a* or *b* in the first place will be $\frac{2}{6}$, suppose one of them taken, as for Instance *a*, then the Probability of taking *b* will be $\frac{1}{5}$. Wherefore the Probability of taking both *a* and *b* will be $\frac{2}{6} \times \frac{1}{5} = \frac{2}{30} = \frac{1}{15}$.

Corollary. The taking of both *a* and *b* is but one single Case of all those by which Six Things may be combined two and two; wherefore the number of Combinations of Six Things taken two and two will be $\frac{6}{1} \times \frac{5}{2} = 15$.

Generally. The number of Combinations of *n* Things, taken two and two, will be $\frac{n}{1} \times \frac{n-1}{2}$.

PROBLEM XIX.

ANY *number of Things* a, b, c, d, e, f *being given: To find the Probability, that in taking three of them as it Happens, they shall be any three proposed, as* a, b, c; *no respect being had to Order.*

SOLUTION.

THE Probability of taking either *a*, or *b*, or *c* in the first place will be $\frac{3}{6}$. Suppose one of them as *a* to be taken, then the Probability of taking *b*, or *c* in the second place will be $\frac{2}{5}$. Again let either of them taken, as suppose *b*; then the Probability of taking *c* in the third place will be $\frac{1}{4}$; wherefore the Probability of taking the three Things proposed, *viz.* *a*, *b*, *c* will be $\frac{3}{6} \times \frac{2}{5} \times \frac{1}{4}$.

Corol-

Corollary. The taking of *a*, *b*, *c* is but one single Cafe of all thofe by which Six Things may be combined three and three; wherefore the number of Combinations of Six Things taken three and three will be $\frac{6}{1} \times \frac{5}{2} \times \frac{4}{3} = 20$.

Generally. The number of Combinations of *n* Things combined according to the number *p*, will be
$$\frac{n \times n-1 \times n-2 \times n-3 \times n-4}{p \times p-1 \times p-2 \times p-3 \times p-4}, \&c.$$ Both Numerator and Denominator being continued to fo many Terms as there are Units in *p*.

PROBLEM XX.

TO *find what Probability there is, that in taking as it Happens Seven Counters out of Twelve, whereof four are White and eight Black, three of them fhall be White ones.*

SOLUTION.

Firft, Find the number of Chances for taking three White ones out of four, which will be $\frac{4}{1} \times \frac{3}{2} \times \frac{2}{3} = 4$.

Secondly, Find the number of Chances for taking four Black ones out of eight: Thefe Chances will be found to be
$$\frac{8 \times 7 \times 6 \times 5}{1 \quad 2 \quad 3 \quad 4} = 70.$$

Thirdly, Becaufe every one of the preceding Chances may be joined with every one of the latter, it follows, that the number of Chances for taking three White ones and four Black ones, will be $4 \times 70 = 280$.

Fourthly, Find the number of Chances for taking four White ones out of four, which will be found to be
$$\frac{4}{1} \times \frac{3}{2} \times \frac{2}{3} \times \frac{1}{4} = 1.$$

Fifthly, Find the number of Chances for taking three Black ones out of eight, which will be $\frac{8}{1} \times \frac{7}{2} \times \frac{6}{3} = 56$.

Sixthly, Multiply thefe two laft numbers together, and the Product 56 will fhew that there are 56 Chances for taking four White ones and three Black ones; which is a Cafe not expreffed in the Problem, yet is implyed: For he who undertakes to take three White Counters out of eight, is reputed to be a winner tho' he takes four; unlefs the contrary be exprefly ftipulated.

Seventhly,

Seventhly, Wherefore the number of Chances for taking three White Counters will be $280 + 56 = 336$.

Eighthly, Seek the number of all the Chances for taking seven Counters out of twelve, which will be found to be $\frac{12 \times 11 \times 10 \times 9 \times 8 \times 7 \times 6}{7 \times 6 \times 5 \times 4 \times 3 \times 2 \times 1} = 792$.

Lastly, Divide the preceding number 336 by the last 792, and the Quotient $\frac{336}{792}$, or $\frac{14}{33}$ will be the Probability required.

Corollary. Let n be the number of all the Counters, a the number of White ones, b the number of Black ones, c the number of Counters to be taken out of the number n; then the number of Chances for taking none of the White ones, or one single White, or two White ones and no more, or three White ones and no more, or four White ones and no more, &c. will be expreſt as follows.

$$\frac{b}{1} \times \frac{b-1}{2} \times \frac{b-2}{3} \times \frac{b-3}{4} \ \&c. \ \times \frac{a}{1} \times \frac{a-1}{2} \times \frac{a-2}{3} \times \frac{a-3}{4} \ \&c.$$

The number of Terms wherein b enters being always equal to $c - a$, and the whole number of Terms equal to c.

But the number of all the Chances for taking a certain number c of Counters out of the number n, with one or more White ones, or without any, will be

$$\frac{n}{1} \times \frac{n-1}{2} \times \frac{n-2}{3} \times \frac{n-3}{4} \times \frac{n-4}{5} \times \frac{n-5}{6} \times \frac{n-6}{7} \times \frac{n-7}{8} \ \&c.$$

which Series muſt be continued to ſo many Terms as there are Units in c.

REMARK.

IF the numbers n and c were large, ſuch as 40000 and 8000, the foregoing method would ſeem impracticable, by reaſon of the vaſt number of Terms to be taken in both Series, whereof the firſt is to be divided by the ſecond: Tho' if thoſe Terms were actually ſet down, a great many of them being common Diviſors, might be expunged out of both Series. However to avoid the trouble of ſetting down ſo many Terms, it will be convenient to uſe the following Theorem, which is a contraction of that Method.

Let therefore n be the number of all the Counters, a the number of White ones, c the number of Counters to be taken

out

out of the number *n*, *p* the number of White Counters to be taken precisely in the number *c*: Then making *n—c = d*, I say that the Probability of taking precisely the number *p* of White Counters will be

$$\frac{\overline{c \times c-1 \times c-2} \; \&c. \quad \overline{\times d \times d-1 \times d-2} \; \&c. \quad \overline{\times \frac{c}{1} \times \frac{c-1}{2} \times \frac{c-2}{3}} \&c.}{n \times n-1 \times n-2 \times n-3 \times n-3 \times n-4 \times n-5 \times n-6 \; \&c.}$$

Here it is to be observed, that the Numerator consists of three Series, which are to be Multiplied together; whereof the first contains as many Terms as there are Units in *p*, the second as many as there are Units in *c—p*, the third as many as there are Units *p*: And the Denominator as many as there are Units in *c*.

PROBLEM XXI.

*I*N *A Lottery consisting of* 40000 *Tickets, among which are Three particular Benefits: What is the Probability that taking* 8000 *of them, one or more of the Three particular Benefits shall be amongst them?*

SOLUTION.

*F*Irst in the Theorem belonging to the Remark of the fore-going Problem, having substituted respectively, 8000, 40000, 32000, 3 and 1, in the room of *c, n, d, c* and *p*; it will appear, that the Probability of taking precisely one of the Three particular Benefits will be

$$\frac{8000 \times 32000 \times 31999 \times 3}{40000 \times 39999 \times 39998}, \text{ or } \frac{48}{125} \text{ nearly.}$$

Secondly, c, n, d, c being interpreted as before, let us sup-pose *p = 2*. Hence the Probability of taking precisely Two of the particular Benefits will be found to be

$$\frac{8000 \times 7999 \times 32000 \times 3}{40000 \times 39999 \times 39998}, \text{ or } \frac{12}{125} \text{ very neat.}$$

Thirdly, Making *p = 3*. The Probability of taking all the Three particular Benefits will be found to be

$$\frac{8000 \times 7999 \times 7998}{40000 \times 39998 \times 39998}, \text{ or } \frac{1}{125} \text{ very near.}$$

P

Where-

Wherefore the Probability of taking one or more of the Three particular Benefits will be $\frac{48+12+1}{125}$, or $\frac{61}{125}$ very near.

N. B. Thefe three Operations might be contracted into one, by inquiring what the Probability is, that none of the particular Benefits may be taken; for then it will be found to be $\frac{32000 \times 31999 \times 31998}{40000 \times 39999 \times 39998} = \frac{64}{125}$ nearly; which being fub-tracted from 1, the Remainder $1 - \frac{64}{125}$, or $\frac{61}{125}$ will fhew the Probability required.

PROBLEM XXII.

T O *Find how many Tickets ought be taken in a Lottery con-fifting of* 40000, *among which there are Three particu-lar Benefits, to make it as Probable that one or more of thofe Three may be taken as not.*

SOLUTION.

L ET the number of Tickets requifite to be taken be x: It will follow therefore from the Theorem belonging to the Remark of the XX*th.* Problem, that the Probability of not taking amongft them any of the particular Benefits will be $\frac{n-x}{n} \times \frac{n-x-1}{n-1} \times \frac{n-x-2}{n-2}$. But this Probability is $= \frac{1}{2}$, fince the Probability of the contrary is $\frac{1}{2}$ by Hypothefis; whence it follows that $\frac{n-x}{1} \times \frac{n-x-1}{2} \times \frac{n-x-2}{3} = \frac{1}{2}$. This Equation being folved, the value of x will be found to be nearly 8252.

N. B. The Factors, whereof both Numerator and Deno-minator are compofed being in Arithmetic Progreffion, and the difference being very fmall in refpect of n; thofe Terms may be confidered as being in Geometric Progreffion, where-fore the Cube of the middle Term $\frac{n-x-1}{n-1}$ may be fuppofed equal to the Product of thofe Terms; from whence will a-rife the Equation $\overline{\frac{n-x-1}{n-1}}^{3} = \frac{1}{2}$ or $\frac{\overline{n-x}^{3}}{n^{3}} = \frac{1}{2}$ (ne-glecting Unity in both Numerator and Denominator) and

con-

consequently x will be found to be $=x \times \overline{1 - \sqrt[3]{\frac{1}{r}}}$, but n is

$= 40000$, and $1 - \sqrt[3]{\frac{1}{r}} = 0.2063$; Therefore $x = 8252$.

In the Remark belonging to the V*th* Problem, a Rule was given for finding the number of Tickets that were to be taken to make it as Probable that one or more of the Benefits fhould be taken, as not; but in that Rule it is fuppofed that the proportion of the Blanks to the Prizes was often repeated, as it ufually is in Lotteries: Now in the Cafe of the prefent Problem, the particular Benefits being but Three in all, the remaining Tickets are to be confidered as Blanks in refpect of them; from whence it follows, that the proportion of the number of Blanks to one Prize being very near as 13332 to 1; and that proportion being repeated but three times in the whole number of Tickets, the Rule there given would not have been fufficiently exact in this Cafe; to fupply which it was thought neceffary to give the Solution of this Problem.

PROBLEM XXIII.

TO *Find at* Pharaon, *how much it is that the Banker gets per Cent of all the Money that is adventured,*

HYPOTHESIS.

I Suppofe, *Firft*, that there is but one fingle Ponte: *Secondly*, That he lays his Money upon one fingle Card at a time: *Thirdly*, That he begins to take a Card in the beginning of the Game : *Fourthly*, That he continues to take a new Card after the laying down of every Pull: *Fifthly*, That when there remains but Six Cards in the Stock, he ceafes to take a Card.

SOLUTION.

WHEN at any time the Ponte lays a new Stake upon a Card taken as it Happens out of his Book, let the number of Cards that are already laid down by the Banker be fuppofed equal to x.

Now

Now in this circumſtance, the Card taken by the Ponte has either paſt four times, or three times, or twice, or once, or not at all.

Firſt, If it has paſſed four times, he can be no loſer upon that account.

Secondly, If it has paſſed three times, then his Card is once in the Stock; now the number of Cards remaining in the Stock being $n-x$, it follows by the firſt Caſe of the XIII*th* Problem that the loſs of the Ponte will be $\frac{1}{n-x}$: But by the Remark belonging to the XX*th* Problem, the Probability that his Card has paſſed three times preciſely in x Card, is $\frac{x \times \overline{x-1} \times \overline{x-2} \times \overline{n-x} \times 4}{n \times \overline{n-1} \times \overline{n-2} \times \overline{n-3}}$. Now ſuppoſing the Denominator equal to s, Multiply the loſs he will ſuffer (if he has that Chance) by the Probability of having it, and the Product $\frac{x \times \overline{x-1} \times \overline{x-2} \times 4}{s}$, will be his abſolute loſs in that circumſtance.

Thirdly, If it has paſſed twice, his loſs by the ſecond Caſe of the XIII*th* Problem will be $\frac{\frac{1}{2}n-\frac{1}{2}x+1}{n-x \times \overline{n-x-1}}$, but the Probability that his Card has paſſed twice in x Cards, is by the Remark of the XX*th* Problem, $\frac{x \times \overline{x-1} \times \overline{n-x} \times \overline{n-x-1} \times 6}{s}$; wherefore Multiplying the loſs he will ſuffer (if he has that Chance) by the Probability of his having it, the Product $\frac{x \times \overline{x-1} \times \frac{1}{2}\overline{n-\frac{1}{2}x+1} \times 6}{s}$ will be his abſolute loſs in that circumſtance.

Fourthly, If it has paſſed once, his loſs Multiplyed into the Probability that it has paſſed, will make his abſolute loſs to be $\frac{x \times \overline{n-x} \times \overline{n-x-2} \times 2}{s}$.

Fifthly, If it has not yet paſſed, his loſs Multiplyed into the Probability that it has not paſſed, will make his abſolute loſs to be $\frac{n-x \times \overline{n-x-1} \times \overline{2n-2x-5}}{s}$.

Now the Sum of all theſe loſſes of the Ponte's will be $\frac{n^3-\frac{1}{2}nn+5n-3x-\frac{3}{2}xx+3x^2}{s}$, and this is the loſs he ſuffers by venturing a new Stake after any number of Cards x are paſt.

But

But the number of Pulls which at any time are laid down, is always one half of the number of Cards that are paſt; wherefore calling t the number of thoſe Pulls, the Loſs of the Ponte may be expreſſed thus, $\dfrac{n^3 - \frac{3}{2}nn + 5n - 6t - 6tt + 24t^3}{t}$.

Let now p be the number of Stakes which the Ponte adventures; let alſo the Loſs of the Ponte be divided into two parts, viz. $\dfrac{n^3 - \frac{3}{2}nn + 5n}{t}$, and $\dfrac{-6t - 6tt + 24t^3}{t}$.

And ſince he adventures a Stake p times; it follows, that the firſt part of his Loſs will be $\dfrac{pn^3 - \frac{3}{2}pnn + 5pn}{t}$.

In order to find the ſecond part, let t be interpreted ſucceſſively by 0, 1, 2, 3 &c. to the laſt Term $p-1$; Then in the room of $6t$ we ſhall have a ſum of numbers in Arithmetic Progreſſion to be Multiplyed by 6; in the room of $6tt$ we ſhall have a ſum of Squares whoſe Roots are in Arithmetic Progreſſion to be Multiplyed by 6; and in the room of $24t^3$ we ſhall have a ſum of Cubes whoſe Roots are in Arithmetic Progreſſion to be Multiplyed by 24: Theſe ſeveral ſums being collected, according to the II*d* Remark on the XI*th* Problem, will be found to be $\dfrac{6p^4 - 14p^3 + 6pp + 2p}{t}$, and therefore the whole Loſs of the Ponte will be

$$\dfrac{pn^3 - \frac{3}{2}pnn + 5pn + 6p^4 - 14p^3 + 6pp + 2p}{t}.$$

Now this being the Loſs which the Ponte ſuſtains by adventuring the ſum p, each Stake being ſuppoſed equal to Unity, it follows, that the Loſs *per Cent* of the Ponte, or the Gain *per Cent* of the Banker is $\dfrac{n^3 - \frac{3}{2}nn + 5n + 6p^3 - 14pp + 6p + 2}{t} \times 100$,

or $\dfrac{2n-5}{2 \times n-1 \times n-3} + \dfrac{p-1 \times 6pp - 8p - 2}{n \times n-1 \times n-2 \times n-3} \times 100$. Let now n be interpreted by 52, and p by 23; and the Gain of the Banker will be found to be 2.99251, that is 2 *l.* 19*sh.* 10*d.* *per Cent.*

By the ſame Method of arguing, it will be found that the Gain *per Cent* of the Banker, at *Baſſete*, will be

$\dfrac{3n-9}{n \times n-1 \times n-2} + \dfrac{4p \times p-1 \times p-2}{n \times n-1 \times n-2 \times n-3} \times 100$. Let n be interpreted by 51, and p by 23; and the foregoing expreſſion

Q   will

will become 0.790582, or 15ˢʰ· 9 *d.* half-penny. The confideration of the firſt Stake, which is adventured before the
Pack is turned, being here omitted as being out of the general Rule: But if that Caſe be taken in, and the Ponte adventures 100 *l.* in 24 Stakes, the Gain of the Banker will be
diminiſhed, and becomes only 0.76245, that is, 15 ˢʰ· 3 *d.* very near: And this is to be eſtimated, as the gain *per Cent* of
the Banker when he takes but half Face.

Now whether the Ponte takes one Card at a time or ſeveral Cards, the Gain *per Cent* of the Banker continues the
ſame: Whether the Ponte keeps conſtantly to the ſame Stake,
or ſome times doubles or triples it, the Gain *per Cent* is ſtill
the ſame: Whether there be but one ſingle Ponte or ſeveral, his Gain *per Cent* is not thereby altered. Wherefore the
Gain *per Cent* of the Banker of all the Money that is adventured at *Pharaon* is 2 *l.* 19ˢʰ· 10 *d.* and at *Baſſete* 15ˢʰ· 3*d.*

PROBLEM XXIV.

SUppoſing A and B *to play together, the Chances they have re-*
ſpectively to win being as a *to* b, *and* B *obliging himſelf to*
Set to A, *ſo long as* A *wins without interruption: What is the*
Advantage that A *gets by his Hand?*

SOLUTION.

Firſt, If *A* and *B* each Stake One, the Gain of *A* on
the firſt Game is $\frac{a-b}{a+b}$.

Secondly, His Gain on the ſecond Game will alſo be $\frac{a-b}{a+b}$,
provided he ſhould happen to win the firſt: But the Probability of *A's* winning the firſt Game is $\frac{a}{a+b}$. Wherefore
his Gain on the ſecond Game will be $\frac{a}{a+b} \times \frac{a-b}{a+b}$.

Thirdly, His Gain on the third Game, after winning the two
firſt, will be likewiſe $\frac{a-b}{a+b}$: But the Probability of *A's* winning
the two firſt Games is $\frac{aa}{a+a}$; Wherefore his Gain on the

third

third Game, when it is eſtimated before the Play begins, is $\frac{aa}{a+b|^2} \times \frac{a-b}{a+b}$ &c.

Fourthly, Wherefore the Gain of the Hand of A is an infinite Series, *viz.* $1 + \frac{a}{a+b} + \frac{aa}{a+b|^2} + \frac{a^3}{a+b|^3} + \frac{a^4}{a+b|^4}$ &c. to be Multiplyed by $\frac{a-b}{a+b}$. But the ſum of that infinite Series is $\frac{a+b}{b}$; Wherefore the Gain of the Hand of A is $\frac{a+b}{b} \times \frac{a-b}{a+b} = \frac{a-b}{b}$.

Corollary I. If A has the advantage of the Odds, and B Sets his Hand out, the Gain of A is the difference of the numbers expreſſing the Odds divided by the leſſer. Thus. if A has the Odds of Five to Three, then his Gain will be $\frac{5-3}{3} = \frac{2}{3}$

Corollary II. If B has the Diſadvantage of the Odds, and A Sets his Hand out, the Loſs of B will be the difference of the number expreſſing the Odds divided by the greater: Thus if B has but Three to Five of the Game, his Loſs will be $\frac{2}{5}$.

Corollary III. If A and B do mutually engage to Set to one-another as long as either of them wins without interruption, the Gain of A will be found to be $\frac{aa-bb}{ab}$: That is the ſum of the numbers expreſſing the Odds Multiplyed by their difference, the product of that Multiplication being divided by the Product of the numbers expreſſing the Odds. Thus if the Odds were as Five to Three, the ſum of 5 and 3. is 8, and the difference 2; Multiply 8 by 2, and the Product 16 being divided by 15 (Product of the number expreſſing the Odds) the Quotient will be $\frac{16}{15}$, or $1\frac{1}{15}$, which therefore will be the Gain of A.

PROBLEM XXV.

ANY *given number of Letters* a, b, c, d, e, f *&c. all of them different, being taken promiſcuouſly, as it Happens: To find the Probability that ſome of them ſhall be found in their places,*
according

according to the rank they obtain in the Alphabet; and that others of them shall at the same time be found out of their places.

SOLUTION.

LET the number of all the Letters be $= n$; let the number of those that are to be in their places be $= p$, and the number of those that are to be out of their places $= q$. Suppose for Brevity sake $\frac{1}{n} = r$, $\frac{1}{n \times n-1} = s$, $\frac{1}{n \times n-1 \times n-2} = t$, $\frac{1}{n \times n-1 \times n-2 \times n-3} = v$ &c. then let all the Quantities $1, r, s, t, v$ &c. be written down with Signs alternately positive and negative, beginning at 1, if p be $= 0$; at r, if $p = 1$; at s, if $p = 2$ &c. Prefix to these Quantities the respective Coefficients of a Binomial Power, whose Index is $= q$: This being done, those Quantities taken all together will express the Probability required; thus the Probability that in Six Letters taken promiscuously, two of them, *viz. a* and *b* shall be in their places, and three of them, *viz. c, d, e* out of their places, will be

$$\frac{1}{6\times5} - \frac{3}{6\times5\times4} + \frac{3}{6\times5\times4\times3} - \frac{1}{6\times5\times4\times3\times2} = \frac{11}{720},$$

And the Probability that *a* shall be in its place, and *b, c, d, e* out of their places, will be

$$\frac{1}{6} - \frac{4}{6\times5} + \frac{6}{6\times5\times4} - \frac{4}{6\times5\times4\times3} + \frac{1}{6\times5\times4\times3\times2} = \frac{53}{720}.$$

The Probability that *a* shall be in its place, and *b, c, d, e, f* out of their places, will be

$$\frac{1}{6} - \frac{5}{6\times5} + \frac{10}{6\times5\times4} - \frac{10}{6\times5\times4\times3} + \frac{5}{6\times5\times4\times3\times2}$$
$$- \frac{1}{6\times5\times4\times3\times2\times1} = \frac{44}{720}, \text{ or } \frac{11}{180}.$$

The Probability that *a, b, c, d, e, f* shall be all displaced is,

$$1 - \frac{6}{6} + \frac{15}{6\times5} - \frac{20}{6\times5\times4} + \frac{15}{6\times5\times4\times3} - \frac{6}{6\times5\times4\times3\times2}$$
$$+ \frac{1}{6\times5\times4\times3\times2\times1}, \text{ or } 1 - 1 + \frac{1}{2} - \frac{1}{6} + \frac{1}{24} - \frac{1}{120}$$
$$+ \frac{1}{720} = \frac{265}{720} = \frac{53}{144}.$$

Hence

Hence it may be concluded that the Probability that one or more of them will be found in their places is $1 - \frac{1}{2} + \frac{1}{6} - \frac{1}{24} + \frac{1}{120} - \frac{1}{720} = \frac{91}{144}$; and that the Odds that one or more of them will be so found are as 91 to 53.

N. B. So many Terms of this last Series are to be taken as there are Units in *n*.

DEMONSTRATION.

THE number of Chances for the Letter *a* to be in the first place contains the number of Chances, by which *a* being in the first place, *b* may be in the second, or out of it: This is an Axiom of common Sense, of the same degree of Evidence as that the Whole is equal to all its Parts.

From this it follows, that if from the number of Chances that there are for *a* to be in the first place, there be subtracted the number of Chances that there are for *a* to be in the first place, and *b* at the same time in the second, there will remain the number of Chances, by which *a* being in the first place, *b* may be excluded the second.

For the same reason it follows, that if from the number of Chances that there are for *a* and *b* to be respectively in the first and second places, there be subtracted the number of Chances by which *a*, *b* and *c* may be respectively in the first, second and third places; there will remain the number of Chances by which *a* being in the first and *b* in the second, *c* may be excluded the third place : And so of the rest.

Let $+ a'$ denote the Probability that *a* shall be in the first place, and let $- a'$ denote the Probability of its being out of it. Likewise let the Probabilities that *b* shall be in the second place or out of it be respectively exprest by $+ b''$ and $- b''$.

Let the Probability that, *a* being in the first place, *b* shall be in the second, be exprest by $a' + b''$: Likewise let the Probability that *a* being in the first place, *b* shall be excluded the second, be exprest by $a' - b''$.

Generally. Let the Probability there is, that as many as are to be in their proper places, shall be so, and at the same time that as many others as are to be out of their proper places

R   shall

shall be so found, be denoted by the particular Probabilities of their being in their proper places, or out of them, written all together: So that for Instance $a' + b'' + c''' - d'''' - e'''''$ may denote the Probability that a, b and c shall be in their proper places, and that at the same time both d and e shall be excluded their proper places.

Now to be able to derive a proper conclusion by vertue of this Notation, it is to be observed, that of the Quantities which are here considered, those from which the Subtraction is to be made, are indifferently composed of any number of Terms connected by + and —; the Quantities which are to be subtracted do exceed by one Term those from which the subtraction is to be made; the rest of the Terms being alike and their signs alike: And the remainder will contain all the Quantities that are alike with their own signs, and also the Quantity Exceeding, but with its sign varied.

It having been demonstrated in what we have said of Permutations and Combinations, that $a' = \frac{1}{n}$, $a' + b'' = \frac{1}{n \times n-1}$, $a' + b'' + c''' = \frac{1}{n \times n-1 \times n-2}$, let $\frac{1}{n}$, $\frac{1}{n \times n-1}$ &c. be respectively called r, s, t, v &c. This being supposed, we may come to the following conclusions.

$$b'' \qquad = r$$
$$b'' + a' = s$$

Therefore $\overline{b'' - a' = r - s}$

$$c'' + b'' \qquad = s \qquad \text{for the same reason that } a' + b'' = s$$
$$c''' + b'' + a' = t$$

2^e. Theref. $\overline{c'' + b'' - a' = \quad s - t}$

$$c''' - a' \qquad = r - s \qquad \text{By the first Conclusion.}$$
$$c''' - a' + b'' = \qquad s - t \qquad \text{By the 2d.}$$

3^e. Theref. $\overline{c''' - a' - b'' = r - 2s + t}$

$$d'''' + c''' + b'' \qquad = t$$
$$d'''' + c''' + b'' + a' = v$$

4^e. Theref. $\overline{d'''' + c''' + b'' - a' = t - v}$

$$d'''' + c''' - a' \qquad = s - t \qquad \text{By the 2d. Conclusion.}$$
$$d'''' + c''' - a' + b'' = \qquad t - v \text{ By the 4th.}$$

5^e. Theref. $\overline{d'''' + c''' - a' - b'' = s - 2t + v}$

$$d''' - b'' - d' \qquad = r - 2s + t \qquad \text{By the 3d. Conc.}$$
$$d'''' - b'' - d' + c'' = \qquad s - 2t + v \qquad \text{By the 5th.}$$

6° Theref. $\overline{d'''' - b'' - d' - c'' = r - 3s + 3t - v}$

By the fame procefs, if no Letter be particularly affigned to be in its place, the Probability that fuch of them as are affigned may be out of their places will likewife be found thus.

$$-a' \qquad = 1 - r \qquad \text{For} + a' \text{ and } -a' \text{ together make}$$
$$-a' + b'' = \qquad r - s \qquad \text{[Unity.}$$

7° Theref. $\overline{-a' - b'' = 1 - 2r + s}$

$$-a' - b'' \qquad = 1 - 2r + s \qquad \text{By the 7th. Conc.}$$
$$-a' - b'' + c'' = \qquad r - 2s + t \qquad \text{By the 3d. Conc.}$$

8° Theref. $\overline{-a' - b'' - c'' = 1 - 3r - 3s - t}$

Now examining carefully all the foregoing Conclufions, it will be perceived, that when the Queftion runs barely upon the difplacing any given number of Letters without requiring that any other fhould be in its place, but leaving it wholly indifferent, then the vulgar Algebraic Quantities which lie on the right hand of the Equations, begin conftantly with Unity: It will alfo be perceived, that when one fingle Letter is affigned to be in its place, then thofe Quantities begin with r; and that when two Letters are affigned to be in their places, they begin with s, and fo on. Moreover 'tis obvious, that thefe Quantities change their figns alternately, and that the Numerical Coefficients which are prefixt to them are thofe of a Binomial Power, whofe Index is equal to the number of Letters which are to be difplaced.

PROBLEM XXVI.

ANY given number of different Letters a, b, c, d, e, f &c. being each of them repeated a certain number of times, and taken promifcuoufly as it Happens: To find the Probability that of fome of thofe Sorts, fome one Letter of each may be found in its proper place, and at the fame time that of fome other Sorts, no one Letter be found in its place.

SOLU-

SOLUTION.

SUppofe *n* be the number of all the Letters, *l* the number of times that each Letter is repeated, and confequently $\frac{n}{l}$ the number of Sorts: Suppofe alfo that *p* be the number of Sorts that are to have one Letter of each in its place; and *q* the number of Sorts of which no one Letter is to be found in its place. Let now the prefcriptions given in the preceding Problem be followed in all refpects, faving that *r* muft here be made $= \frac{l}{n}$, $s = \frac{l^1}{n \times n - 1}$, $t = \frac{l^1}{n \times n - 1 \times n - 2}$ &c. and the Solution of any particular Cafe of the Problem will be obtained.

Thus if it were required to find the Probability that no Letter of any fort fhall be in its place, the Probability thereof would be

$$1 - qr + \frac{q}{1} \times \frac{q-1}{2} s - \frac{q}{1} \times \frac{q-1}{2} \times \frac{q-2}{3} t \ \&c.$$

But in this particular Cafe *q* would be equal to $\frac{n}{l}$, wherefore the foregoing Series might be changed into this, *viz.*

$$\frac{1}{2} \times \overline{\frac{n-l}{n-1}} - \frac{1}{6} \times \overline{\frac{n-l \times n - 2l}{n-1 \times n-2}} + \frac{1}{24} \times \overline{\frac{n-l \times n - 2l \times n - 3l}{n-1 \times n-2 \times n - 3}}$$
&c.

Corollary I. From hence it follows, that the Probability that one or more Letters indeterminately taken may be in their places will be

$$1 - \frac{1}{2} \times \overline{\frac{n-l}{n-1}} + \frac{1}{6} \times \overline{\frac{n-l \times n - 2l}{n-1 \times n-2}} - \frac{1}{24} \times \overline{\frac{n-l \times n-2l \times n-3l}{n-1 \times n-2 \times n-3}}$$
&c.

Corollary II. The Probability that two or more Letters indeterminately taken may be in their places will be expreft as follows,

$$\frac{1}{4} \times \overline{\frac{n-l}{n-1}} - \frac{2}{113} \times \overline{\frac{n-2l}{n-2}} A + \frac{3}{214} \times \overline{\frac{n-3l}{n-3}} B - \frac{4}{315} \times \overline{\frac{n-4l}{n-4}} C + \frac{5}{416} \times \overline{\frac{n-5l}{n-5}} D \ \&c.$$

Corollary III. The Probability that three or more Letters indeterminately taken may be in their places will be as follows,

$$\tfrac{1}{6} \times \tfrac{\overline{n-1} \times \overline{n-2l}}{\overline{n-1} \times \overline{n-2}} - \tfrac{3}{1\times4} \times \tfrac{\overline{n-3l}}{\overline{n-3}} A + \tfrac{4}{2\times5} \times \tfrac{\overline{n-4l}}{\overline{n-4}} B$$

$$- \tfrac{5}{3\times6} \times \tfrac{\overline{n-5l}}{\overline{n-5}} C + \tfrac{6}{4\times7} \times \tfrac{\overline{n-6l}}{\overline{n-6}} D \ \&c.$$

Corollary IV. The Probability that four or more Letters, indeterminately taken, may be in their places will be thus expreſt,

$$\tfrac{1}{24} \times \tfrac{\overline{n-l}}{\overline{n-1}} \times \tfrac{\overline{n-2l}}{\overline{n-2}} \times \tfrac{\overline{n-3l}}{\overline{n-3}} - \tfrac{4}{1\times5} \times \tfrac{\overline{n-4l}}{\overline{n-4}} A + \tfrac{5}{2\times6} \times \tfrac{\overline{n-5l}}{\overline{n-5}} B$$

$$- \tfrac{6}{3\times7} \times \tfrac{\overline{n-6l}}{\overline{n-7}} C \ \&c.$$

The Law of the continuation of theſe Series being mani-feſt, it will be eaſy to reduce them all to one general Se-ries.

From what we have ſaid it follows, that in a common Pack of 52 Cards, the Probability that one of the four Aces may be in the firſt place; one of the four Duces in the ſecond; or one of the four Traes in the third; or that ſome one of any other ſort may be in its place (making 13 different places in all) will be expreſt by the Series exhibited in the firſt Corollary.

It follows likewiſe, that if there be two Packs of Cards, and that the Order of the Cards in one of the Packs be the Rule whereby to eſtimate the rank which the Cards of the ſame Suite and Name are to obtain in the other; the Pro-bability that one Card or more, in one of the Packs, may be found in the ſame Poſition as the like Card in the other Pack, will be expreſt by the Series belonging to the firſt Corol-lary, making $n = 52$ and $l = 1$: Which Series will in this Caſe be $1 - \tfrac{1}{2} + \tfrac{1}{6} - \tfrac{1}{24} + \tfrac{1}{120} - \tfrac{1}{720}$ &c. whereof 52 Terms ought to be taken.

If the Terms of the foregoing Series are joined by couples, the Series will become,

$$\tfrac{1}{2} + \tfrac{1}{2\times4} + \tfrac{1}{2\times3\times4\times6} + \tfrac{1}{2\times3\times4\times5\times6\times8} + \tfrac{1}{2\times3\times4\times5\times6\times7\times8\times10}$$

&c. of which 26 Terms ought to be taken.

But by reaſon of the great Convergency of the aforeſaid Series, a few of its Terms will give a ſufficient approxima-

S   tion

tion in all Cases required; as appears by the following Operation,

$$\frac{1}{2} = 0.500000$$

$$\frac{1}{2\cdot4} = 0.125000.$$

$$\frac{1}{2\cdot3\cdot4\cdot6} = 0.006944 +$$

$$\frac{1}{2\cdot3\cdot4\cdot5\cdot6\cdot8} = 0.000174 +$$

$$\frac{1}{2\cdot3\cdot4\cdot5\cdot6\cdot7\cdot8\cdot10} = 0.000002 +$$

$$\text{Sum} = 0.632129 +$$

Wherefore the Probability that one or more like Cards in two different Packs may obtain the same Position, will be in all Cases very near 0.632; and the Odds that this will Happen once or oftner, as 632 to 368, or as 12 to 7 very near.

But the Odds that two or more like Cards in two different Packs will not obtain the same Position, are very nearly as 736 to 264 or 14 to 5.

Corollary V. If *A* and *B*, each holding a Pack of Cards, pull them out at the same time one after another, on condition that every time two like Cards are pulled out, *A* shall give *B* a Guinea; and it were required to find what consideration *B* ought to give *A* to Play on those terms: The Answer will be, One Guinea, let the number of Cards be what it will.

Corollary VI. If the number of Packs be given, the Probability that any given number of circumstances may Happen in them all, or in any of them, will be found easily by our method. Thus, if the number of the Packs be *k*, the Probability that one Card or more of the same Sute and Name, in every one of the Packs, may be in the same Position, will be exprest as follows.

$$\frac{1}{2^{k-2}} - \frac{1}{2\cdot\overline{2-1}|^{k-2}} + \frac{1}{6\cdot\overline{2-1\cdot2-2}|^{k-2}} - \frac{1}{24\cdot\overline{2-1\cdot2-2\cdot2-3}|^{k-2}}$$

&c.

PRO-

PROBLEM XXVII.

IF A *and* B *play together, each with a certain number of Bowls* = *n*: *What are their respective Probabilities of winning, supposing that each of them want a certain number of Games of being up?*

SOLUTION.

First, the Probability that some Bowl of *B* may be nearer the Jack than any Bowl of *A* is $\frac{1}{2}$.

Secondly, Supposing one of his Bowls nearer the Jack than any Bowl of *A*, the number of his remaining Bowls is *n*—1, and the number of all the Bowls remaining between them is 2*n*—1: Wherefore the Probability that some other of his Bowls may be nearer the Jack than any Bowl of *A* will be $\frac{n-1}{2n-1}$, from whence it follows, that the Probability of his winning two Bowls or more is $\frac{1}{2} \times \frac{n-1}{2n-1}$.

Thirdly, Supposing two of his Bowls nearer the Jack than any Bowl of *A*, the Probability that some other of his Bowls may be nearer the Jack than any Bowl of *A* will be $\frac{n-2}{2n-2}$: Wherefore the Probability of winning three Bowls or more is $\frac{1}{2} \times \frac{n-1}{2n-1} \times \frac{n-2}{2n-2}$: The continuation of which process is manifest.

Fourthly, The Probability that one single Bowl of *B* shall be nearer the Jack than any Bowl of *A* is $\frac{1}{2} - \frac{1}{2} \times \frac{n-1}{2n-1}$, or $\frac{1}{2} \times \frac{n}{2n-1}$; For, if from the Probability that one or more of his Bowls may be nearer the Jack than any Bowl of *A*, there be subtracted the Probability that two or more may be nearer, there remains the Probability of one single Bowl of *B* being nearer: In this Case *B* is said to Win one Bowl at an End.

Fifthly, The Probability that two Bowls of *B*, and not more, may be nearer the Jack than any Bowl of *A*, will be found to be $\frac{1}{2} \times \frac{n-1}{2n-1} \times \frac{n}{2n-2}$, in which Case *B* is said to win two Bowls at an End.

Sixthly,

Sixthly, The Probability that B may win three Bowls at an End will be found to be $\frac{1}{2} \times \frac{n-1}{2n-1} \times \frac{n-2}{2n-2} \times \frac{n}{2n-3}$. The procefs whereof is manifeft.

The Reader may obferve, that the foregoing Expreffions might be reduced to fewer Terms; but leaving them unreduced, the Law of the procefs is thereby made more confpicuous.

Let it carefully be obferv'd, when we mention henceforth the Probability of winning two Bowls, that the Senfe of it ought to be extended to two Bowls or more; and that when we mention the winning two Bowls at an End, it ought to be taken in the common acceptation of two Bowls only: The like being to be obferved in other Cafes.

This Preparation being made; fuppofe, *Firft*, that A wants one Game of being up, and B two; and let it be required, in that circumftance, to determine their Probabilities of winning.

Let the whole Stake between them be fuppofed $= 1$. Then either A may win a Bowl, or B win one Bowl at an End, or B may win two Bowls.

In the firft Cafe B lofes his Expectation.

In the fecond Cafe he becomes intitled to $\frac{1}{2}$ of the Stake. But the Probability of this Cafe is $\frac{1}{2} \times \frac{n}{2n-1}$: wherefore his Expectation arifing from that part of the Stake he will be intitled to, if this Cafe fhould Happen, and from the Probability of its Happening, will be $\frac{1}{4} \times \frac{n}{2n-1}$.

In the third Cafe B wins the whole Stake 1. But the Probability of this Cafe is $\frac{1}{2} \times \frac{n-1}{2n-1}$: wherefore the Expectation of B upon that account is $\frac{1}{2} \times \frac{n-1}{2n-1}$.

From this it follows that the whole Expectation of B is $\frac{1}{4} \times \frac{n}{2n-1} + \frac{1}{2} \times \frac{n-1}{2n-1}$ or $\frac{\frac{3}{4}n - \frac{1}{2}}{2n-1}$, or $\frac{3n-2}{8n-4}$; which being fubtracted from Unity, the remainder will be the Expectation of A, *viz.* $\frac{5n-2}{8n-4}$. It may therefore be concluded, that the Probabilities which A and B have of winning are refpectively as $5n-2$ to $3n-2$.

'Tis remarkable, that the fewer the Bowls are, the greater is the proportion of the Odds; for if A and B play with
fingle

fingle Bowls, the proportion will be as 3 to 1; if they play with two Bowls each, the proportion will be as 2 to 1; if with three Bowls each, the proportion will be as 13 to 7: yet let the number of Bowls be never fo great, that proportion will not defcend fo low as 5 to 3.

Secondly, Suppofe *A* wants one Game of being up, and *B* three; then either *A* may win a Bowl, or *B* win one Bowl at an End, or two Bowls at an End, or three Bowls.

In the firft Cafe *B* lofes his Expectation.

If the fecond Cafe Happens, then *B* will be in the circumftance of wanting but two to *A*'s one; in which Cafe his Expectation will be $\frac{3n-2}{6n-4}$, as it has been before determined: but the Probability that this Cafe may Happen is $\frac{1}{2} \times \frac{n}{2n-1}$; wherefore the Expectation of *B*, arifing from the profpect of this Cafe, will be $\frac{1}{2} \times \frac{n}{2n-1} \times \frac{3n-2}{6n-4}$.

If the third Cafe Happen, then *B* will be intitled to one half of the Stake: but the Probability of its Happening is $\frac{1}{2} \times \frac{n-1}{2n-1} \times \frac{n}{2n-2}$; wherefore the Expectation of *B* arifing from the Profpect of this Cafe is $\frac{1}{4} \times \frac{n-1}{2n-1} \times \frac{n}{2n-2}$, or $\frac{1}{8} \times \frac{n}{2n-1}$.

If the fourth Cafe Happen, then *B* wins the whole Stake 1: but the Probability of its Happening is $\frac{1}{2} \times \frac{n-1}{2n-1} \times \frac{n-2}{2n-2}$, or $\frac{1}{4} \times \frac{n-2}{2n-1}$; wherefore the Expectation of *B* arifing from the profpect of this Cafe will be found to be $\frac{1}{4} \times \frac{n-2}{2n-1}$.

From this it follows, that the whole Expectation of *B* will be $\frac{9nn-13n+4}{8 \times \overline{2n-1}|^2}$; which being fubtracted from Unity, the remainder will be the Expectation of *A*, *viz.* $\frac{23nn-19n+4}{8 \times \overline{2n-1}|^2}$.

It may therefore be concluded, that the Probabilities which *A* and *B* have of winning are refpectively as $23nn-19n+4$ to $9nn-13n+4$.

N. B. If *A* and *B* play only with One Bowl each, the Expectation of *B* deduced from the foregoing Theorem would be found $= 0$. which we know from other principles ought to be $= \frac{1}{8}$. The reafon of which is that the Cafe of winning Two Bowls at an End, and the Cafe of winning

Three

Three Bowls at an End, enter this conclusion, which Cases do not belong to the supposition of playing with single Bowls: wherefore excluding those two Cases, the Expectation of *B* will be found to be $\frac{1}{2} \times \frac{2}{2n-1} \times \frac{3n-2}{6n-4} = \frac{1}{6}$, which will appear if *n* be made $= 1$. Yet the Expectation of *B*, in the Case of two Bowls, would be rightly determined, tho' the Case of winning Three Bowls at an End enters it: The reason of which is, that the Probability of winning Three Bowls at an End is $=, \frac{1}{4} \times \frac{n-2}{2n-1}$, which in the Case of Two Bowls becomes $= 0$, so that the general Expression is not thereby disturbed.

After what we have said, it will be easy to extend this way of Reasoning to any circumstance of Games wanting between *A* and *B*; by making the Solution of each simpler Case subservient to the Solution of that which is immediately more compound.

Having given formerly the Solution of this Problem, proposed to me by the Honourable *Frances Robarts*, in the *Philosophical Transactions* Number 339; I there said, by way of Corollary, that if the proportion of Skill in the Gamesters were given, the Problem might also be Solved; since which time *Mr de Monmort*, in the second Edition of a Book by him Published upon the subject of Chance, has thought it worth his while to Solve this Problem as it is extended to the consideration of the Skill, and to carry his Solution to a very great number of Cases, giving also a Method by which it may still be carried farther: I very willingly acknowledge his Solution to be extreamly good, and own that he has in this, as well as in a great many other things, shewn himself entirely master of the doctrine of Combinations, which he has employed with very great Industry and Sagacity.

The Solution of this Problem, as it is restrained to an equality of Skill, was in my *Specimen* deduced from the Method of Combinations; but the Solution which is given of it in this place, is deduced from a Principle which has more of simplicity in it, being that by the help of which I have Demonstrated the Doctrine of Permutations and Combinations: Wherefore to make it as familiar as possible, and to shew its vast extent, I shall now apply it to the general Solution

of

of this Problem, taking in the confideration of the *Skill* of the Gamefters.

But before I proceed I think it neceffary to define what I call Skill: *viz.* That it is the proportion of Chances which the Gamefters may be fuppofed to have for winning a fingle Game with one Bowl each.

PROBLEM XXVIII.

IF A and B, whofe proportion of Skill is as a *to* b, *play together, each with a certain number of Bowls: What are their refpective Probabilities of winning, fuppofing each of them to want a certain number of Games of being up?*

SOLUTION.

Firft, The Chance of *B* for winning one fingle Bowl being *b*, and the number of his Bowls being *n*, it follows that the fum of all his Chances is *nb*; and for the fame reafon the fum of all the Chances of *A* is *na*: wherefore the fum of all the Chances for winning one Bowl or more is *na + nb*; which for brevity fake we may call *s*. From whence it follows, that the Probability which *B* has of winning one Bowl or more is $\frac{nb}{s}$.

Secondly, Suppofing one of his Bowls nearer the Jack than any of the Bowls of *A*, the number of his remaining Chances is $\overline{n-1} \times b$; and the number of Chances remaining between them is *s — b*: wherefore the Probability that fome other of his Bowls may be nearer the Jack than any Bowl of *A* will be $\frac{\overline{n-1} \times b}{s-b}$: From whence it follows, that the Probability of his winning Two Bowls or more is $\frac{nb}{s} \times \frac{\overline{n-1} \times b}{s-b}$.

Thirdly, Suppofing Two of his Bowls nearer the Jack than any of the Bowls of *A*, the number of his remaining Chances is $\overline{n-2} \times b$; and the number of Chances remaining between them is *s — 2 b*: wherefore the Probability that fome other of his Bowls may be nearer the Jack than any Bowl of *A* will be $\frac{\overline{n-2} \times b}{s-2b}$. From whence it follows, that the

Proba-

Probability of his winning Three Bowls or more is $\frac{nb}{s} \times$ $\frac{\overline{n-1} \times b}{s-b} \times \frac{\overline{n-2} \times b}{s-2b}$; the continuation of which proceſs is manifeſt.

Fourthly, If from the Probability which B has of winning One Bowl or more, there be ſubtracted the Probability which he has of winning Two or more, there will remain the Probability of his winning One Bowl at an End: Which therefore will be found to be $\frac{nb}{s} - \frac{nb}{s} \times \frac{\overline{n-1} \times b}{s-b}$ or $\frac{nb}{s} \times \frac{s-nb}{s-b}$ or $\frac{nb}{s} \times \frac{sa}{s-b}$.

Fifthly, For the ſame reaſon as above, the Probability which B has of winning Two Bowls at an End will be found to be $\frac{nb}{s} \times \frac{\overline{n-1} \times b}{s-b} \times \frac{sa}{s-2b}$.

Sixthly, And for the ſame reaſon likewiſe, the Probability which B has of winning Three Bowls at an End will be found to be $\frac{nb}{s} \times \frac{\overline{n-1} \times b}{s-b} \times \frac{\overline{n-2} \times b}{s-2b} \times \frac{sa}{s-3b}$. The continuation of which proceſs is manifeſt.

N. B. The ſame Expectations which denote the Probability of any circumſtance of B, will denote likewiſe the Probability of the like circumſtance of A, only changing b into a and a into b.

Theſe Things being premiſed, Suppoſe *Firſt*, that each of them wants one Game of being up; 'tis plain that the Expectations of A and B are reſpectively $\frac{aa}{s}$ and $\frac{ba}{s}$. Let this Expectation of B be called P.

Secondly, Suppoſe A wants One Game of being up and B Two, and let the Expectation of B be required: Then either A may win a Bowl, or B win One Bowl at an End, or B win Two Bowls.

If the firſt Caſe Happens, B loſes his Expectation.

If the ſecond Happens, he gets the Expectation P; but the Probability of this Caſe is $\frac{nb}{s} \times \frac{sa}{s-b}$: wherefore the Expectation of B ariſing from the poſſibility that it may ſo Happen is $\frac{nb}{s} \times \frac{sa}{s-b} \times P$.

If the third Cafe Happens, he gets the whole Stake 1; but the Probability of this Cafe is $\frac{nb}{s} \times \frac{s-b}{s-b}$, wherefore the Expectation of B arifing from the Probability of this Cafe is $\frac{nb}{s} \times \frac{s-b}{s-b} \times 1$.

From which it follows that the whole Expectation of B will be $\frac{nb}{s} \times \frac{sn}{s-b} P + \frac{nb}{s} \times \frac{s-b}{s-b}$. Let this Expectation be called Q.

Thirdly, Suppofe A to want One Game of being up, and B Three. Then either B may win One Bowl at an End, in which Cafe he gets the Expectation Q; or Two Bowls at an End, in which Cafe he gets the Expectation P; or Three Bowls in which Cafe he gets the whole Stake 1. Wherefore the Expectation of B will be found to be $\frac{nb}{s} \times \frac{sn}{s-b} \times Q$

$+ \frac{nb}{s} \times \frac{\overline{n-1} \times b}{s-b} \times \frac{sn}{s-2b} \times P + \frac{nb}{s} \times \frac{\overline{n-1} \times b}{s-b} \times \frac{\overline{n-2} \times b}{s-2b}.$

An infinite number of thefe Theorems may be formed in the fame manner, which may be continued by infpection, having well obferved how each of them is deduced from the preceding.

If the number of Bowls were unequal, fo that A had m Bowls and B n Bowls; Suppofing $ma + nb = s$, other Theorems might be found to anfwer that inequality: And if that inequality fhould not be conftant, but vary at pleafure; other Theorems might alfo be formed to anfwer that Variation of inequality, by following the fame way of arguing. And if Three or more Gamefters were to play together under any circumftance of Games wanting, and of any given proportion of Skill, their Probabilities of winning might be determined after the fame manner.

PROBLEM XXIX.

TO find the Expectation of A when with a Die of any given number of Faces he undertakes to fling any determinate number of them in any given number of Cafts.

U   SOLU-

SOLUTION.

LET $p+1$ be the number of all the Faces in the Die, n the number of Casts, f the number of Faces which he undertakes to fling.

The number of Chances for an Ace to come up once or more in any number of Casts n, is $\overline{p+1}\,^n - p^n$: As has been proved in the Introduction.

Let the Duce, by thought, be expunged out of the Die, and thereby the number of its Faces reduced to p, then the number of Chances for the Ace to come up will at the same time be reduced to $p^n - \overline{p-1}\,^n$. Let now the Duce be restored, and the number of Chances for the Ace to come up without the Duce, will be the same as if the Duce were expunged. But if from the number of Chances for the Ace to come up with or without the Duce, *viz.* from $\overline{p+1}\,^n - p^n$ be subtracted the number of Chances for the Ace to come up without the Duce, *viz.* $p^n - \overline{p-1}\,^n$, there will remain the number of Chances for the Ace and Duce to come up once or more, which consequently will be $\overline{p+1}\,^n - 2 \times p^n + \overline{p-1}\,^n$.

By the same way of arguing it will be proved, that the number of Chances for the Ace and Duce to come up without the Trae will be $p^n - 2 \times \overline{p-1}\,^n + \overline{p-2}\,^n$, and consequently, that the number of Chances for the Ace, the Duce and Trae to come up once or more, will be the difference between $\overline{p+1}\,^n - 2 \times p^n + \overline{p-1}\,^n$ and $p^n - 2 \times \overline{p-1}\,^n + \overline{p-2}\,^n$; which therefore is $\overline{p+1}\,^n - 3 \times p^n + 3 \times \overline{p-1}\,^n + \overline{p-2}\,^n$.

Again it may be proved that the number of Chances for the Ace, the Duce, the Trae and Quater to come up, is $\overline{p+1}\,^n - 4 \times p^n + 6 \times \overline{p-1}\,^n - 4 \times \overline{p-2}\,^n + \overline{p-3}\,^n$; the continuation of which Process is manifest.

Wherefore if all the Powers $\overline{p+1}\,^n, p^n, \overline{p-1}\,^n, \overline{p-2}\,^n, \overline{p-3}\,^n$ &c. with the Signs alternately Positive and Negative, be written in Order, and to those Powers there be prefixt the respective Coefficients of a Binomial raised to the Power f; the sum of all those Terms will be the Numerator of the Expectation of A, of which the Denominator will be $\overline{p+1}\,^n$.

E X A M-

EXAMPLE I.

LET Six be the number of Faces in the Die, and let *A* undertake in Eight Casts to fling both an Ace and a Duce: Then his Expectation will be $\frac{6^8 - 2 \times 5^8 + 4^8}{6^8}$

$= \frac{964502}{1680216} = \frac{4}{7}$ nearly.

EXAMPLE II.

IF *A* undertake with a common Die to fling all the Faces in 12 Casts, his Expectation will be found to be

$\frac{6^{12} - 6 \times 5^{12} + 15 \times 4^{12} - 20 \times 3^{12} + 15 \times 2^{12} - 6 \times 1^{12}}{6^{12}} = \frac{10}{23}$

nearly.

EXAMPLE III.

IF *A* with a Die of 36 Faces undertake to fling two given Faces in 43 Casts; or, which is the same thing, if with two common Dice he undertake in 43 Casts to fling Two Aces at one time, and Two Sixes at another time, his Expectation will be $\frac{36^{43} - 2 \times 35^{43} + 34^{43}}{36^{43}} = \frac{49}{100}$ nearly.

N. B. The parts of which these Expectations are compounded, are easily obtained by the help of a Table of Logarithms.

PROBLEM XXX.

TO *find in how many Trials it will be probable that* A *with a Die of any given number of Faces shall throw any proposed number of them.*

SOLUTION.

LET $p + 1$ be the number of Faces in the Die, and f the number of Faces which are to be thrown. Divide the Logarithm of $\frac{1}{1 - \sqrt[f]{\frac{1}{2}}}$ by the Logarithm of $\frac{p+1}{p}$, and the

the Quotient will exprefs nearly the number of Trials requifite, to make it as probable that the propofed Faces may be
thrown as not.

DEMONSTRATION.

SUppofe Six to be the number of Faces which are to be
thrown, and n the number of Trials: Then by what has
been demonftrated in the preceding Problem, the Expectation of A will be,

$$\frac{\overline{p+1}^{n}-6\times p^{n}+15\times\overline{p-1}^{n}-20\times\overline{p-2}^{n}+15\times\overline{p-3}^{n}-6\times\overline{p-4}^{n}+\overline{p-5}^{n}}{\overline{p+1}^{n}}$$

Let it be fuppofed that the Terms $\overline{p+1}$, p, $p-1$, $p-2$ &c.
are in Geometric Progreffion (which fuppofition will
very little err from the truth, efpecially if the proportion of p to 1 be not very fmall). Let now r be written inftead of $\frac{p+1}{p}$, and then the Expectation of A will be
changed into $1-\frac{6}{r^{n}}+\frac{15}{r^{2n}}-\frac{20}{r^{3n}}+\frac{15}{r^{4n}}-\frac{6}{r^{5n}}+\frac{1}{r^{6n}}$,
or $\overline{1-\frac{1}{r^{n}}}^{6}$. But this Expectation of A ought to be made
equal to $\frac{1}{2}$, fince by fuppofition he has an equal Chance to
win or lofe: Hence will arife the Equation $\overline{1-\frac{1}{r^{n}}}^{6}=\frac{1}{2}$
or $r^{n}=\frac{1}{1-\sqrt[6]{\frac{1}{2}}}$, from which it may be concluded that
$n\times$ Log. r, or $n\times$ Log. $\frac{p+1}{p}=$ Log. $\frac{1}{1-\sqrt[6]{\frac{1}{2}}}$, and confequently that n is equal to the Logarithm of $\frac{1}{1-\sqrt[6]{\frac{1}{2}}}$ divided by the
Logarithm of $\frac{6}{5}=\frac{p+1}{p}$. And the fame Demonftration will
hold in any other Cafe.

EXAMPLE I.

TO find in how many Trials A may with equal Chance
undertake to throw all the Faces of a common Die.

The

The Logarithm of $\frac{1}{1-\sqrt[6]{\frac{1}{4}}}$ = 0.9621753; the Loga-

rithm of $\frac{p+1}{p}$ or $\frac{6}{5}$ = 0.0791812: Wherefore n

= $\frac{0.9621753}{0.0791812}$ = 12 +. From hence it may be concluded
that in 12 Casts A has the worst of the Lay, and in 13
the best of it.

EXAMPLE II.

TO find in how many Trials, A may with equal Chance,
with a Die of Thirty-six Faces, undertake to throw
Six determinate Faces; or, in how many Trials he may with
a Pair of common Dice undertake to throw all the Doublets.

The Logarithm of $\frac{1}{1-\sqrt[6]{\frac{1}{4}}}$ being 0.9621753, and the

Logarithm of $\frac{p+1}{p}$ or $\frac{36}{35}$ being 0.0122345; it follows

that the number of Casts requisite to that effect is $\frac{0.9621753}{0.0122345}$

or 79 nearly.

But if it were the Law of the Play, that the Doublets
must be thrown in a given Order, and that any Doublet
Happening to be thrown out of its turn should go for nothing; then the throwing of the Six Doublets would be like
the throwing of the two Aces Six times; to produce which
effect the number of Casts requisite would be found by Multiplying 35 by 5.668, as appears from our VII*th*. Problem,
and consequently would be about 198.

N. B. the Fraction $\frac{1}{1-\sqrt[p]{\frac{1}{4}}}$ may be reduced to $\frac{\sqrt[p]{2}}{\sqrt[p]{2}-1}$

which will Facilitate the taking of its Logarithm.

PROBLEM XXXI.

IF A, B, C *Play together on the following conditions; First, that
they shall each of them Stake* 1 l. *Secondly, that A and B shall
begin the Play; Thirdly, that the Loser shall yield his place to the
third Man, which is to be observed constantly afterwards; Fourthly,*

X

that

that the Loſer ſhall be fined a certain Sum p, which is to ſerve to increaſe the common Stock; Laſtly, that he ſhall Win the whole Sum depoſited at firſt, and increaſed by the ſeveral Fines, who ſhall firſt beat the other two ſucceſſively: 'Tis demanded what is the Advantage of A and B, whom we ſuppoſe to begin the Play.

SOLUTION.

LEt *B A* ſignifie that *B* beats *A*, and *A C* that *A* beats *C*; and let always the firſt Letter denote the Winner, and the ſecond the Loſer.

Let us ſuppoſe that *B* beats *A* the firſt time: Then let us inquire what the Probability is that the Set ſhall be ended in any given number of Games; and alſo what is the Probability which each Gameſter has of winning the Set in that given number of Games.

Firſt, If the Set be ended in two Games, *B* muſt neceſſarily be the winner; for by Hypotheſis he wins the firſt time: Which may be expreſſed as follows.

$$\begin{array}{c|c} 1 & B\,A \\ 2 & B\,C \end{array}$$

Secondly, If the Set be ended in Three Games, *C* muſt be the winner; as appears by the following Scheme.

$$\begin{array}{c|c} 1 & B\,A \\ 2 & C\,B \\ 3 & C\,A \end{array}$$

Thirdly, If the Set be ended in Four Games, *A* muſt be the winner; as appears by this Scheme.

$$\begin{array}{c|c} 1 & B\,A \\ 2 & C\,B \\ 3 & A\,C \\ 4 & A\,B \end{array}$$

Fourthly, If the Set be ended in Five Games, *B* muſt be the winner; which is thus expreſſed,

$$\begin{array}{c|c} 1 & B\,A \end{array}$$

1	BA
2	CB
3	AC
4	BA
5	BC

Fifthly, If the Set be ended in Six Games, C muſt be the winner; as will appear by ſtill following the ſame Proceſs, thus,

1	BA
2	CB
3	AC
4	BA
5	CB
6	CA

And this Proceſs recurring continually in the ſame Order needs not be proſecuted any farther.

Now the Probability that the firſt Scheme ſhall take place is $\frac{1}{2}$, in conſequence of the ſuppoſition that B beats A the firſt time; it being an equal Chance whether B beat C, or C beat A.

And the Probability that the ſecond Scheme ſhall take place is $\frac{1}{4}$: For the Probability of C beating B is $\frac{1}{2}$, and that being ſuppoſed, the Probability of his beating A will alſo be $\frac{1}{2}$; wherefore the Probability of B beating C, and then A, will be $\frac{1}{2} \times \frac{1}{2}$ or $\frac{1}{4}$.

And from the ſame conſiderations the Probability that the Third Scheme ſhall take place is $\frac{1}{8}$: and ſo on.

Hence it will be eaſie to compoſe a Table of the Probabilities which B, C, A have of winning the Set in any given number of Games; and alſo of their Expectations: Which Expectations are the Probabilities of winning Multiplyed by the Stock Three depoſited at firſt, and increaſed ſuccceſſively by the ſeveral Fines.

TABLE

TABLE *of the Probabilities,* &c.

	B	C	A
2	$\frac{1}{2} \times \overline{3+2p}$		
3		$\frac{1}{4} \times \overline{3+3p}$	
4			$\frac{1}{8} \times \overline{3+4p}$
5	$\frac{1}{16} \times \overline{3+5p}$		
6		$\frac{1}{32} \times \overline{3+6p}$	
7			$\frac{1}{64} \times \overline{3+7p}$
8	$\frac{1}{128} \times \overline{3+8p}$		
9		$\frac{1}{256} \times \overline{3+9p}$	
10			$\frac{1}{512} \times \overline{3+10p}$
11	$\frac{1}{1024} \times \overline{3+11p}$		
&c.			

Now the feveral Expectations of *B, C, A* may be fummed up by the following Lemma.

LEMMA.

$$\frac{n}{b} + \frac{n+d}{bb} + \frac{n+2d}{b^3} + \frac{n+3d}{b^4} + \frac{n+4d}{b^5} \&c. \ Ad\ infinitum$$

is equal to $\frac{n}{b-1} + \frac{d}{\overline{b-1}^2}$.

Let the Expectations of *B* be divided into two Series, *viz.*

$$\frac{3}{2} + \frac{3}{16} + \frac{3}{128} + \frac{3}{1024} \ \&c.$$
$$+ \frac{2p}{2} + \frac{5p}{16} + \frac{8p}{128} + \frac{11p}{1024} \ \&c.$$

The firft Series conftitutes a Geometric Progreffion continually decreafing, whofe fum will be found to be $\frac{12}{7}$.

The fecond Series may be reduced to the form of the Series in our Lemma, and may be thus expreft,

$$\frac{p}{2} \times$$

$\frac{p}{2} \times \frac{2}{1} + \frac{5}{8} + \frac{8}{8^2} + \frac{11}{8^3} + \frac{14}{8^4}$ &c. Wherefore
dividing the whole by $\frac{p}{2}$, and laying aside the Term 2, we
shall have the Series $\frac{5}{8} + \frac{8}{8^2} + \frac{11}{8^3} + \frac{14}{8^4}$, &c. which
has the same form as the Series of the Lemma, and may be
compared with it : Let therefore n be made $= 5$, $d = 3$ and
$b = 8$, and the sum of this Series will be $\frac{1}{7} + \frac{3}{49}$, or
$\frac{3^2}{49}$; to this adding the first Term 2, which had been laid
aside, the new sum will be $\frac{136}{49}$; and that being Multiplied by $\frac{p}{2}$, the Product will be $\frac{68}{49} p$, which is the sum
of the second Series expressing the Expectations of B: From
hence it may be concluded, that all the Expectations of B
contained in both the abovementioned Series will be equal
to $\frac{12}{7} + \frac{68}{49} p$.

And by the help of the foregoing Lemma it will be
found likewise that all the Expectations of C will be equal
to $\frac{6}{7} + \frac{48}{49} p$.

It will also be found that all the Expectations of A will
be $= \frac{3}{7} + \frac{31}{49} p$.

Hitherto we have determined the several Expectations of
the Gamesters, upon the sum by them deposited at first, as
also upon the Fines by which the common Stock is increased: It remains now to Estimate the several Risks of their
being Fined; that is to say, the sum of the Probabilities of
their being Fined multiplyed by the respective Quantities of
the Fine.

Now after the supposition made of A being beat the first
time, by which he is obliged to lay down his Fine p, B and C
have an equal Chance of being Fined after the second Game,
which makes the Risk of each to be $= \frac{1}{2} p$, as appears
by the following Scheme.

$$\frac{BA}{CB} \quad \text{or} \quad \frac{BA}{BC}$$

Y

In

In the like manner, it will be found that both C and A have one Chance in four for their being Fined after the Third Game, and confequently that the Risk of each is $\frac{1}{4}p$, according to the following Scheme.

$$\frac{BA}{CB} \text{ or } \frac{BA}{CB}$$
$$AC \qquad CA$$

And by the like Procefs it will be found that the Risk of B and C after the fourth Game is $\frac{1}{8}p$.

Hence it will be eafie to compofe the following Table which exprefles the Risks of each Gamefter.

TABLE *of* RISKS.

	B	C	A
2	$\frac{1}{2}p$	$\frac{1}{2}p$	- - - - - - -
3	- - - - - - -	$\frac{1}{4}p$	$\frac{1}{4}p$
4	$\frac{1}{8}p$	- - - - - - -	$\frac{1}{8}p$
5	$\frac{1}{16}p$	$\frac{1}{16}p$	- - - - - - -
6	- - - - - - -	$\frac{1}{32}p$	$\frac{1}{32}p$
7	$\frac{1}{64}p$	- - - - - - -	$\frac{1}{64}p$
8	$\frac{1}{128}p$	$\frac{1}{128}p$	- - - - - - -
9	- - - - - - -	$\frac{1}{256}p$	$\frac{1}{256}p$
&c.			

In the Column belonging to B, if the vacant places were filled up, and the Terms $\frac{1}{4}p$, $\frac{1}{32}p$, $\frac{1}{256}p$ &c. were Interpoled, the Sum of the Risks of B would compofe one uninterrupted Geometric Progreffion, whofe Sum would be $= p$; But the Terms interpoled conftitute a Geometric Progreffion whofe Sum is $= \frac{2}{7}p$: Wherefore, if from p there be fubtraĉted $\frac{2}{7}p$, there will remain $\frac{5}{7}p$ for the Sum of the Risks of B.

In like manner it will be found that the Sum of the Risks of C will be $= \frac{6}{7}p$.   And

And the Sum of the Risks of *A*, after his being Fined the firſt time, will be $= \frac{3}{7} p$.

Now if from the ſeveral Expectations of the Gameſters there be ſubtracted each Man's Stake, as alſo the Sum of his Risks, there will remain the clear Gain or Loſs of each of them.

Wherefore, from the Expectations of $B = \frac{12}{7} + \frac{63}{49} p$
Subtracting *firſt* his Stake $\qquad = 1$
Then the Sum of his Risks $\qquad = \qquad \frac{5}{7} p$
_____
There remains the clear Gain of $B = \frac{5}{7} + \frac{33}{49} p$
_____

Likewiſe, from the Expectations of $C = \frac{6}{7} + \frac{48}{49} p$
Subtracting *firſt* his Stake $\qquad = 1$
Then the Sum of his Risks $\qquad = \qquad \frac{6}{7} p$
_____
There remains the clear Gain of $C = -\frac{1}{7} + \frac{6}{49} p$
_____

In like manner, from the Expectation of $A = \qquad \frac{3}{7} + \frac{21}{49} p$
Subtracting, *Firſt*, his Stake $\qquad = 1$
Secondly, the Sum of his Risks $= \qquad \frac{3}{7} p$
Laſtly, the Fine *p* due to the ⎱ $=$
Stock by the Loſs of the firſt Game ⎰ $\qquad\qquad p$
_____
There remains the clear Gain of $A = -\frac{4}{7} - \frac{20}{49} p$.
_____

But we have ſuppoſed in the beginning of the Game that *A* was beat; whereas *A* had the ſame Chance to beat *B*, as *B* had to beat him: Wherefore dividing the Sum of the Gains of *B* and *A* into two equal Parts, each part will be $\frac{1}{14} - \frac{2}{49} p$; which conſequently muſt be reputed to be as the Gain of each of them.

Corollary I. The Gain of *C* being $-\frac{1}{7} + \frac{6}{49} p$, Let that be made $= 0$. Then *p* will be found $= \frac{7}{6}$. If therefore the Fine has the ſame proportion to each Man's Stake as 7 has to 6, the Gameſters play all upon equal Terms: But if the Fine bears a leſs proportion to the Stake than 7
to .

to 6, C has the difadvantage: Thus, Suppofing $p = 1$, his Lofs would be $\frac{1}{49}$. But if the Fine bears a greater proportion to the Stake than 7 to 6, C has the Advantage.

Corollary II. If the Stake were conftant, that is, if there were no Fines, then the Probabilities of winning would be refpectively proportional to the Expectations; wherefore fuppoling $p = 0$, the Expectations of the Gamefters, or their Probabilities of winning, will be as $\frac{12}{7}$, $\frac{6}{7}$, $\frac{3}{7}$, or, as 4, 2, 1 : But the increafe of the Stock caufes no alteration in the Probabilities of winning, and confequently thofe Probabilities are, in the Cafe of this Problem, as 4, 2, 1; whereof the firft belongs to B after his beating A the firft time; the fecond to C, and the third to A : Wherefore 'tis Five to Two, before the Play begins, that either A or B wins the Set; and Five to Four that one of them, that fhall be fixt upon, wins it.

Corollary III. If the proportion of Skill between the Gamefters A, B, C be as a, b, c refpectively, and that the refpective Probabilities of winning, in any number of Games after the firft, wherein B is Suppofed to beat A, be denoted by B', B'''', B''', B^X &c, C'', C^V, C^{VIII}, C^{XI} &c. A''', A^{VI}, A'^X, $A^{X_{II}}$, &c. it will be found, by the bare infpection of the Schemes belonging to the Solution of the foregoing Problem, that

$$B' = \frac{b}{b+c}$$

$$C'' = \frac{c}{c+b} \times \frac{c}{c+a}$$

$$A'' = \frac{c}{c+b} \times \frac{a}{a+c} \times \frac{a}{a+b}$$

$$B''' = \frac{c}{c+b} \times \frac{a}{a+c} \times \frac{b}{b+a} \times \frac{b}{b+c}$$

$$C^V = \frac{c}{c+b} \times \frac{a}{a+c} \times \frac{b}{b+a} \times \frac{c}{c+b} \times \frac{c}{c+a}$$

$$A^{V} = \frac{c}{c+b} \times \frac{a}{a+c} \times \frac{b}{b+a} \times \frac{c}{c+b} \times \frac{a}{a+c} \times \frac{a}{a+b}.$$
&c.

Let $\frac{b}{b+a} \times \frac{c}{c+b} \times \frac{a}{a+c}$ be made $= m$; then it will plainly appear that the feveral Probabilities of winning will compofe each of them a Geometric Progreffion, for

$$B''' =$$

$$B''' = m \; B' \quad \Big| \quad C^{v} = m \; C'' \quad \Big| \quad A^{v} = m \; A''$$
$$B^{v''} = m \; B''' \quad \Big| \quad C^{v''} = m \; C^{v} \quad \Big| \quad A^{'x} = m \; A^{v}$$
$$B^{x} = m \; B^{v''} \quad \Big| \quad C^{x'} = m \; C^{v''} \quad \Big| \quad A^{x''} = m \; A^{'x}$$
$$\&c. \qquad\qquad \&c. \qquad\qquad \&c.$$

Hence a Table of Expectations and Risks may eafily be formed as above; and the reft of the Solution carried on by following exactly the fteps of the former.

When the Solution is brought to its conclufion, it will be neceffary to make an allowance for the fuppofition made that *B* beats *A* the firft time, which may be done thus,

Let *P* be the Gain of *B*, when expreft by the Quantities *a, b, c,* and *Q* the Gain of *A*, when expreft by the fame: Change *a* into *b* and *b* into *a,* in the Quantity *Q;* then the Quantity refulting from this Change will be the Gain of *B,* in cafe he be fuppofed to lofe the firft Game. Let this Quantity therefore be called *R*, and then the Gain of *B*, to be eftimated before the Play begins, will be $\frac{bP + aR}{b + a}$.

PROBLEM XXXII.

IF *Four Gamefters* A, B, C, D *Play on the conditions of the foregoing Problem, and he be to be reputed the Winner, who fhall beat the other Three fucceffively: What is the Advantage of* A *and* B, *whom we Suppofe to begin the Play?*

SOLUTION.

LET *BA* denote, as in the preceding Problem, that *B* beats *A*, and *AC* that *A* beats *C*; and generally let the firft Letter always denote the Winner and the fecond the Lofer.

Let it be Suppofed alfo that *B* beats *A* the firft time: Then let it be inquired what is the Probability that the Play fhall be ended in any given number of Games; as alfo what is the Probability which each Gamefter has of winning the Set in that given number of Games.

Firft,

Firſt, If the Set be ended in Three Games, *B* muſt necef-
farily be the winner: Since by Hypotheſis he beats *A* the
firſt Game, which is expreſſed as follows,

$$
\begin{array}{c|c}
1 & B A \\
2 & B C \\
3 & B D
\end{array}
$$

Secondly, If the Set be ended in Four Games, *C* muſt be
the winner; as it thus appears.

$$
\begin{array}{c|c}
1 & B A. \\
2 & C B. \\
3 & C D \\
4 & C A
\end{array}
$$

Thirdly, If the Set be ended in Five Games, *D* will be
the winner; for which he has two Chances, as it appears by
the following Scheme.

$$
\begin{array}{c|c}
1 & B A \\
2 & C B \\
3 & D C \quad \text{or} \\
4 & D A \\
5 & D B
\end{array}
\qquad
\begin{array}{c}
B A \\
B C \\
D B. \\
D A \\
D C.
\end{array}
$$

Fourthly, If the Set be ended in Six Games, *A* will be
the winner; and he has three Chances for it, which are thus
collected,

$$
\begin{array}{c|c}
1 & B A \\
2 & C B \\
3 & D C \\
4 & A D \\
5 & A B \\
6 & A C
\end{array}
\qquad
\begin{array}{c}
B A \\
C B \\
C D \\
A C. \\
A B \\
A D.
\end{array}
\qquad
\begin{array}{c}
B A \\
B C. \\
D B. \\
A D. \\
A C \\
A B.
\end{array}
$$

Fifthly, If the Set be ended in Seven Games, then *B* will
have three Chances to be the winner, and *C* will have two;
thus,

$$
\begin{array}{c|c}
1 & B A.
\end{array}
$$

1	BA	BA	BA	BA	BA
2	CB	CB	CB	BC	BC
3	DC	DC	CD	DB	DB
4	AD	DA	AC	AD	DA
5	BA	BD	BA	CA	CD
6	BC	BC	BD	CB	CB
7	BD	BA	BC	CD	CA

Sixthly, If the Set be ended in Eight Games, then *D* will have two Chances to be the Winner, *C* will have three, and *B* also three, thus

1	BA	BA	BA	BA	BA	BA	BA	BA
2	CB	CB	CB	CB	CB	BC	BC	BC
3	DC	DC	DC	CD	CD	DB	DB	DB
4	AD	AD	DA	AC	AC	AD	AD	DA
5	BA	AB	BD	BA	AB	CA	AC	CD
6	CB	CA	CB	DB	DA	BC	BA	BC
7	CD	CD	CA	DC	DC	BD	BD	BA
8	CA	CB	CD	DA	DB	BA	BC	BD

Let now the Letters by which the winners are denoted be written in Order, prefixing to them the Numbers which, expreſs their ſeveral Chances for winning; in this manner,

3	1 B.
4	1 C
5	2 D
6	3 A
7	3 B + 2 C
8	3 C + 2 D + 3 B
9	3 D + 2 A + 3 C + 3 D + 2 A
10	3 A + 2 B + 3 D + 3 A + 2 B + 3 A + 2 C + 3 D

&c.

Then

Then Examining the formation of thefe Letters, it will appear; *Firſt*, that the Letter *B* is always found ſo many times in any Rank, as the Letter *A* is found in the two preceding Ranks: *Secondly*, that *C* is found ſo many times in any Rank, as *B* is found in the preceding Rank, and *D* in the Rank before that. *Thirdly*, that *D* is found ſo many times in each Rank, as *C* is found in the preceding, and *B* in the Rank before that: And *Fourthly*, that *A* is found ſo many times in each, as *D* is found in the preceding Rank, and *C* in the Rank before that.

From whence it may be concluded, that the Probability which the Gameſter *B* has of winning the Set, in any given number of Games, is $\frac{1}{2}$ of the Probability which *A* has of winning it one Game ſooner, together with $\frac{1}{4}$ of the Probability which *A* has of winning it two Games ſooner.

The Probability which *C* has of winning the Set, in any given number of Games, is $\frac{1}{2}$ of the Probability which *B* has of winning it one Game ſooner, together with $\frac{1}{4}$ of the Probability which *D* has of winning it two Games ſooner.

The Probability which *D* has of winning the Set, in any given number of Games, is $\frac{1}{2}$ of the Probability which *C* has of winning it one Game ſooner, and alſo $\frac{1}{4}$ of the Probability which *B* has of winning it two Games ſooner.

The Probability which *A* has of winning the Set, in any given number of Games, is $\frac{1}{2}$ of the Probability which *D* has of winning it one Game ſooner, and alſo $\frac{1}{4}$ of the Probability which *C* has of winning it two Games ſooner.

Theſe things being obſerved, it will be eaſie to compoſe a Table of the Probabilities which *B, C, D, A* have of winning the Set in any given number of Games; as alſo of their Expeƈtations, which will be as follows.

A T A-

TABLE *of the Probabilities,* &c.

		B	C	D	A
'	3	$\frac{1}{4} \times \overline{4+3p}$	- - - - - -	- - - - - -	- - - - - -
"	4	- - - - - -	$\frac{1}{8} \times \overline{4+4p}$	- - - - - -	- - - - - -
'''	5	- - - - - -	- - - - - -	$\frac{2}{16} \times \overline{4+5p}$	- - - - - -
''''	6	- - - - - -	- - - - - -	- - - - - -	$\frac{3}{32} \times \overline{4+6p}$
v	7	$\frac{3}{64} \times \overline{4+7p}$	$\frac{2}{64} \times \overline{4+7p}$	- - - - - -	- - - - - -
iv	8	$\frac{3}{128} \times \overline{4+8p}$	$\frac{3}{128} \times \overline{4+8p}$	$\frac{2}{128} \times \overline{4+8p}$	- - - - - -
vii	9	- - - - - -	$\frac{2}{256} \times \overline{4+9p}$	$\frac{6}{256} \times \overline{4+9p}$	$\frac{3}{256} \times \overline{4+9p}$
viii	10	$\frac{1}{512} \times \overline{4+10p}$	$\frac{3}{512} \times \overline{4+10p}$	$\frac{6}{512} \times \overline{4+10p}$	$\frac{3}{512} \times \overline{4+10p}$
'x	11	$\frac{13}{1024} \times \overline{4+11p}$	$\frac{10}{1024} \times \overline{4+11p}$	$\frac{2}{1024} \times \overline{4+11p}$	$\frac{2}{1024} \times \overline{4+11p}$
x	12	$\frac{18}{2048} \times \overline{4+12p}$	$\frac{19}{2048} \times \overline{4+12p}$	$\frac{14}{2048} \times \overline{4+12p}$	$\frac{4}{2048} \times \overline{4+12p}$
&c.	&c.				

The Terms whereof each Column of this Table is compofed, being not eafily fummable by any of the known Methods, it will be convenient, in order to find their Sums, to ufe the following *Analyfis.*

Let $B' + B'' + B''' + B'''' + B^v + B^{vi}$ &c. reprefent the refpective Probabilities which B has of winning the Set, in any number of Games, anfwering to 3, 4, 5, 6, 7, 8 &c. and let the fum of the Probabilities *Ad infinitum* be fuppofed $= y$.

In the fame manner, let $C' + C'' + C''' + C'''' + C^v + C^{vi}$ &c. reprefent the Probabilities which C has of winning, which fuppofe $= z$.

Let the like Probabilities which D has of winning be reprefented by $D' + D'' + D''' + D'''' + D^v + D^{vi}$ &c. which fuppofe $= v$.

Laftly, Let the Probabilities which A has of winning be reprefented by $A' + A'' + A''' + A'''' + A^v + A^{vi}$ &c. which fuppofe $= x$.

Now from the Obfervations fet down before the Table of Probabilities, it will follow, that

A 2 $B' =$

$$B' = B'$$
$$B'' = B''$$
$$B''' = \tfrac{1}{2} A'' + \tfrac{1}{4} A'$$
$$B'''' = \tfrac{1}{2} A''' + \tfrac{1}{4} A''$$
$$B^{v} = \tfrac{1}{2} A'''' + \tfrac{1}{4} A'''$$
$$B^{v'} = \tfrac{1}{2} A^{v} + \tfrac{1}{4} A''''$$
&c.

From which Scheme we may deduce the Equation following, $y = \tfrac{1}{4} + \tfrac{3}{4} x$: For the Sum of the Terms in the firſt Column is equal to the Sum of the Terms in the other two. But the Sum of the Terms in the firſt Column is y by Hypotheſis; wherefore y ought to be made equal to the Sum of the Terms in the other two Columns.

In order to find the Sum of the Terms of the ſecond Column, I argue thus,

$A' + A'' + A''' + A'''' + A^{v} + A^{v'}$ is $= x$ by Hypoth.

or $A'' + A''' + A'''' + A^{v} + A^{v'}$ is $= x - A'$

and $\tfrac{1}{2}A'' + \tfrac{1}{2}A''' + \tfrac{1}{2}A'''' + \tfrac{1}{2}A^{v} + \tfrac{1}{2}A^{v'}$ is $= \tfrac{1}{2}x - \tfrac{1}{2}A'$

Then adding $B' + B''$ on both ſides of the laſt Equation, we ſhall have

$$B' + B'' + \tfrac{1}{2}A'' + \tfrac{1}{2}A''' + \tfrac{1}{2}A'''' + \tfrac{1}{2}A^{v} + \tfrac{1}{2}A^{v'} \text{ &c.}$$
$$= \tfrac{1}{2}x - \tfrac{1}{2}A' + B' + B''.$$

But $A' = 0$, $B' = \tfrac{1}{4}$, $B'' = 0$, as appears from the Table: Wherefore the Sum of the Terms of the ſecond Column is equal to $\tfrac{1}{2}x + \tfrac{1}{4}$.

The Sum of the Terms of the third Column is $\tfrac{1}{4}x$ by Hypotheſis; and conſequently the Sum of the Terms in the ſecond and third Columns is $= \tfrac{3}{4}x + \tfrac{1}{4}$. From whence it follows that the Equation $y = \tfrac{1}{4} + \tfrac{3}{4}x$ had been rightly determined.

In

In the fame manner, if we write

$$C' = C'$$
$$C'' = C''$$
$$C''' = \tfrac{1}{2} B'' + \tfrac{1}{4} D'$$
$$C'''' = \tfrac{1}{2} B''' + \tfrac{1}{4} D''$$
$$C^v = \tfrac{1}{2} B'''' + \tfrac{1}{4} D'''$$
$$C^{vi} = \tfrac{1}{2} B^v + \tfrac{1}{4} D''''$$
&c.

By a reafoning like the former we fhall at length come at the Equation $z = \tfrac{1}{2} y + \tfrac{1}{4} v$.

So likewife if we write

$$D' = D'$$
$$D'' = D''$$
$$D''' = \tfrac{1}{2} C'' + \tfrac{1}{4} B'$$
$$D'''' = \tfrac{1}{2} C''' + \tfrac{1}{4} B''$$
$$D^v = \tfrac{1}{2} C'''' + \tfrac{1}{4} B'''$$
$$D^{vi} = \tfrac{1}{2} C^v + \tfrac{1}{4} B''''$$
&c.

We fhall deduce the Equation $v = \tfrac{1}{2} z + \tfrac{1}{4} y$.

Laftly, if after the fame manner we write

$$A' = A'$$
$$A'' = A''$$
$$A''' = \tfrac{1}{2} D'' + \tfrac{1}{4} C'$$
$$A'''' = \tfrac{1}{2} D''' + \tfrac{1}{4} C''$$
$$A^v = \tfrac{1}{2} D'''' + \tfrac{1}{4} C'''$$
$$A^{vi} = \tfrac{1}{2} D^v + \tfrac{1}{4} C''''$$
&c.

We fhall obtain the Equation $x = \tfrac{1}{2} v + \tfrac{1}{4} z$.

Now,

Now thefe Four Equations being refolved, it will be found that

$$B' + B'' + B''' + B'''' + B^V + B^{VI} \&c. = y = \frac{46}{149},$$
$$C' + C'' + C''' + C'''' + C^V + C^{VI} \&c. = z = \frac{36}{149},$$
$$D' + D'' + D''' + D'''' + D^V + D^{VI} \&c. = v = \frac{32}{149},$$
$$A' + A'' + A''' + A'''' + A^V + A^{VI} \&c. = x = \frac{25}{149}.$$

Thefe Values being once found, let b, c, d, a, which are commonly employed to denote known Quantities, be refpectively fubftituted in the room of them; to the end that the Letters y, z, v, x may now be employed to denote other unknown Quantities.

Hitherto we have been determining the Probabilities of winning: But in order to find the Expectations of the Gamefters, each Term of the Series expreffing thefe Probabilities, is to be multiplyed by the refpective Terms of the following Series; $4+3p, 4+4p, 4+5p, 4+6p, \&c.$

The firft part of each Product being no more than a Multiplication by 4, the fums of all the firft parts of thofe Products are only the fums of the Probabilities multiplied by 4; and confequently are $4b, 4c, 4d,$ and $4a$ refpectively.

But to find the Sums of the other parts,

Let
$$3\ B'p + 4\ B''p + 5\ B'''p + 6\ B''''p \&c. \text{ be } = p\,y,$$
$$3\ C'p + 4\ C''p + 5\ C'''p + 6\ C''''p \&c. \quad = p\,z,$$
$$3\ D'p + 4\ D''p + 5\ D'''p + 6\ D''''p \&c. \quad = p\,v,$$
$$3\ A'p + 4\ A''p + 5\ A'''p + 6\ A''''p \&c. \quad = p\,x,$$

Now Since
$$3\ B' = 3\ B'$$
$$4\ B'' = 4\ B''$$
$$5\ B''' = \tfrac{5}{2} A'' + \tfrac{5}{4} A$$
$$6\ B'''' = \tfrac{6}{2} A''' + \tfrac{6}{4} A'$$
$$7\ B^V = \tfrac{7}{2} A'''' + \tfrac{7}{4} A'''$$
$$8\ B^{VI} = \tfrac{8}{2} A^V + \tfrac{8}{4} A''''$$
$$\&c.$$

It

It follows, that $y = \frac{1}{4} + \frac{3}{4}x + a.$ For the firſt Column is $= y$, by *Hypotheſis.*

Again, $3 A' + 4 A'' + 5 A''' + 6 A'''' + 7 A^v$ &c. $= x$ by *Hy-*
potheſis.

But $A' + A'' + A''' + A'''' + A^v$ &c. has been found $= a,$

Wherefore adding theſe two Equations together, we ſhall
have $4 A' + 5 A'' + 6 A''' + 7 A'''' + 8 A^v$ &c. $= x + a.$
or $\frac{4}{2}A' + \frac{5}{2}A'' + \frac{6}{2}A''' + \frac{7}{2}A'''' + \frac{8}{2}A^v$ &c. $= \frac{1}{2}x + \frac{1}{2}a.$

Now the Terms of this laſt Series, together with $3 B' +$
$4 B''$, compoſe the ſecond Column: But $3 B' = \frac{3}{4}$, and
$4 B'' = 0$, as appears from the Table. Conſequently the
ſum of the Terms of the ſecond Column is $= \frac{3}{4} + \frac{1}{2}x$
$+ \frac{1}{2}a.$

By the ſame Method of proceding, it will be found, that
the ſum of the Terms of the third Column is $= \frac{1}{4}x$
$+ \frac{1}{2}a.$

From whence it follows that $y = \frac{3}{4} + \frac{1}{2}x + \frac{1}{4}a + \frac{1}{4}x + \frac{1}{2}a,$
or $y = \frac{3}{4} + \frac{3}{4}x + a.$

In the ſame manner if we write

$$3\ C' = 3\ C'$$
$$4\ C'' = 4\ C''$$
$$5\ C''' = \frac{5}{2}B'' + \frac{5}{4}D'$$
$$6\ C'''' = \frac{6}{2}C''' + \frac{6}{4}D''$$
$$7\ C^v = \frac{7}{2}B''' + \frac{7}{4}D'''$$
$$8\ C^{vi} = \frac{8}{2}B^v + \frac{8}{4}D''''$$
&c.

We ſhall from thence deduce the Equation $z = \frac{1}{2}y + \frac{1}{2}b$
$+ \frac{1}{4}v + \frac{1}{2}d.$

So likewife in the fame manner, if we write

$$3\ D' = 3\ D'$$
$$4\ D'' = 4\ D''$$
$$5\ D''' = \tfrac{5}{2}\ C'' + \tfrac{5}{4}\ B'$$
$$6\ D'''' = \tfrac{6}{2}\ C''' + \tfrac{6}{4}\ B''$$
$$7\ D^{v} = \tfrac{7}{2}\ D''' + \tfrac{7}{4}\ B'''$$
$$8\ D^{v?} = \tfrac{8}{2}\ D^{v} + \tfrac{8}{4}\ B''''$$
&c.

Laftly, if after the fame manner we write

$$3\ A' = 3\ A'$$
$$4\ A'' = 4\ A''$$
$$5\ A''' = \tfrac{5}{2}\ D' + \tfrac{5}{4}\ C'$$
$$6\ A'''' = \tfrac{6}{2}\ D''' + \tfrac{6}{4}\ C''$$
$$7\ A^{v} = \tfrac{7}{2}\ D''' + \tfrac{7}{4}\ C'''$$
$$8\ A^{v?} = \tfrac{8}{2}\ D^{v} + \tfrac{8}{4}\ C''''$$
&c.

We fhall deduce the two following Equations, *viz.*

$$v = \tfrac{1}{2}z + \tfrac{1}{2}c + \tfrac{1}{4}y + \tfrac{1}{2}b. \text{ And } x = \tfrac{1}{2}v + \tfrac{1}{2}d + \tfrac{1}{4}z + \tfrac{1}{2}c.$$

Now the foregoing Equations being Solved, and the values of *b, c, d, a* reftored, it will be found that $y = \frac{45536}{22201}$, $z = \frac{38724}{22201}$, $v = \frac{37600}{22201}$, $x = \frac{33547}{22201}$.

From which we may conclude, that the feveral Expectations of *B, C, D, A* are refpectively, *Firft,* $4 \times \frac{56}{149} + \frac{45536}{22201} p.$ *Secondly,* $4 \times \frac{36}{149} + \frac{38724}{22201} p.$ *Thirdly,* $4 \times \frac{32}{149} + \frac{37600}{22201} p.$ *Fourthly,* $4 \times \frac{25}{149} + \frac{33547}{22201} p.$

The

The Expectations of the Gamesters being thus found, it will be necessary to find the Risks of their being Fined, or otherwise what sum each of them ought justly to give to have their Fines Insured. In order to which, let us form so many Schemes for expressing the Probabilities of the Fines as are sufficient to find the Law of their Process.

And *First*, we may observe, that upon the supposition of *B* beating *A* the first Game, in consequence of which *A* is to be Fined, *B* and *C* have one Chance each for being Fined the second Game, as it thus appears·

$$
\begin{array}{c|cc}
1 & BA & BA \\
\hline
2 & CB & BC
\end{array}
$$

Secondly, that *C* has one Chance in four for being Fined the third Game, *B* one Chance likewise, and *D* two; according to the following Scheme,

$$
\begin{array}{c|cccc}
1· & BA & BA & BA & BA \\
2 & CB & CB & BC & BC \\
3· & DC & CD & DB & BD
\end{array}
$$

Thirdly, that *D* has two Chances in eight for being Fined the fourth Game, that *A* has three and *C* one; according to the following Scheme,·

$$
\begin{array}{c|cccccc}
1 & BA & BA & BA & BA & BA & BA \\
2 & CB & CB & CB & CB & BC & BC \\
3 & DC & DC & CD & CD & DB & DB \\
4 & AD & DA & AC & CA & AD & DA
\end{array}
$$

N. B. The two Combinations *BA*, *BC*, *BD*, *AB*, and *BA*, *BC*, *BD*, *BA* are omitted in this Scheme, as being superfluous; their disposition shewing that the Set must have been ended in three Games, and consequently not affecting the Gamesters as to the Probability of their being Fined the fourth Game; Yet the number of all the Chances must be reckoned as being Eight; since the Probability of any one circumstance is but $\frac{1}{8}$.

These

Thefe Schemes being continued, it will eafily be perceived that the circumftances under which the Gamefters find themfelves, in refpect of their Risks of being Fined, ftand related to one another in the fame manner as were their Probabilities of Winning; from which confideration a Table of the Risks may eafily be compofed as follows.

A TABLE *of* RISKS &c.

		B	C	D	A
'	2	$\frac{1}{2}P$	$\frac{1}{2}P$	- - - - -	- - - - -
''	3	$\frac{1}{4}P$	$\frac{1}{4}P$	$\frac{2}{4}P$	- - - - -
'''	4	- - - - -	$\frac{1}{8}P$	$\frac{2}{8}P$	$\frac{3}{8}P$
''''	5	$\frac{3}{16}P$	$\frac{2}{16}P$	$\frac{2}{16}P$	$\frac{3}{16}P$
v	6	$\frac{6}{32}P$	$\frac{5}{32}P$	$\frac{2}{32}P$	$\frac{3}{32}P$
ɪᵛ	7	$\frac{6}{64}P$	$\frac{8}{64}P$	$\frac{8}{64}P$	$\frac{4}{32}P$
vɪɪɪ	8	$\frac{7}{128}P$	$\frac{8}{128}P$	$\frac{14}{128}P$	$\frac{13}{128}P$
vɪɪɪ	9	$\frac{17}{256}P$	$\frac{15}{256}P$	$\frac{14}{256}P$	$\frac{22}{256}P$
&c.	&c.				

Wherefore fuppofing $B' + B'' + B'''$ &c. $C' + C'' + C'''$ &c. $D' + D'' + D'''$ &c. $A' + A'' + A'''$ &c. to reprefent the feveral Probabilities; and fuppofing that the feveral fums of thefe Probabilities are refpectively equal to y, x, z, v, we fhall have the following Schemes and Equations

$$B' = B'$$
$$B'' = B''$$
$$B''' = \tfrac{1}{2}A'' + \tfrac{1}{4}A'$$
$$B'''' = \tfrac{1}{2}A''' + \tfrac{1}{4}A''$$
$$B^v = \tfrac{1}{2}A'''' + \tfrac{1}{4}A'''$$
$$B^{ɪᵛ} = \tfrac{1}{2}A^v + \tfrac{1}{4}A''''$$
&c.

Hence $y = \tfrac{3}{4} + \tfrac{1}{4}x$.

$$C' =$$

$$C' = C'$$
$$C'' = C''$$
$$C''' = \tfrac{1}{2}B'' + \tfrac{1}{4}D'$$
$$C'''' = \tfrac{1}{2}B''' + \tfrac{1}{4}D''$$
$$C^v = \tfrac{1}{2}B'''' + \tfrac{1}{4}D'''$$
$$C^{v'} = \tfrac{1}{2}B^v + \tfrac{1}{4}D''''$$
&c.

Hence $z = \tfrac{1}{2} + \tfrac{1}{2}y + \tfrac{1}{4}v$.

$$D' = D'$$
$$D'' = D''$$
$$D''' = \tfrac{1}{2}C'' + \tfrac{1}{4}B'$$
$$D'''' = \tfrac{1}{2}C''' + \tfrac{1}{4}B''$$
$$D^v = \tfrac{1}{2}C'''' + \tfrac{1}{4}B'''$$
$$D^{17} = \tfrac{1}{2}C^v + \tfrac{1}{4}B''''$$
&c.

Hence $v = \tfrac{1}{4} + \tfrac{1}{2}z + \tfrac{1}{4}y$.

$$A' = A'$$
$$A'' = A''$$
$$A''' = \tfrac{1}{2}D'' + \tfrac{1}{4}C'$$
$$A'''' = \tfrac{1}{2}D''' + \tfrac{1}{4}C''$$
$$A^v = \tfrac{1}{2}D'''' + \tfrac{1}{4}C'''$$
$$A^{v'} = \tfrac{1}{2}D^v + \tfrac{1}{4}C''''$$
&c.

Hence $x = \tfrac{1}{2}v + \tfrac{1}{4}z$.

C c

The

The foregoing Equations being refolved, we fhall have

$$y = \tfrac{243}{149}, \ z = \tfrac{252}{149}, \ v = \tfrac{224}{149}, \ x = \tfrac{175}{149},$$

Let every one of thofe Fractions be now multiplied by p, and the Products $\tfrac{243}{149}p, \ \tfrac{252}{149}p, \ \tfrac{224}{149}p, \ \tfrac{175}{149}p$ will exprefs the refpective Risks of B, C, D, A, or the fums they might juftly give to have their Fines Infured.

But if from the feveral Expectations of the Gamefters there be fubtracted, *Firft*, the fums by them depofited in the beginning of the Play, and *Secondly*, the Risks of their Fines, there will remain the clear Gain or Lofs of each. Wherefore

From the Expectations of $B = \quad 4 \times \tfrac{56}{149} + \tfrac{45516}{22201}p$,

Subtracting his own Stake $= \quad 1$

and alfo the fum of the Risks $= \qquad\qquad \tfrac{243}{149}p$,

There remains his clear Gain $= \qquad \tfrac{75}{149} + \tfrac{9329}{22201}p.$

From the Expectations of $C = \quad 4 \times \tfrac{36}{149} + \tfrac{28724}{22201}p$,

Subtracting his own Stake $= \quad 1$

and alfo the Sum of his Risks $= \qquad\qquad \tfrac{252}{149}p$,

There remains his clear Gain $= - \tfrac{5}{149} + \tfrac{1176}{22201}p.$

From the Expectations of $D = \quad 4 \times \tfrac{32}{149} + \tfrac{37600}{22201}p$,

Subtracting his own Stake $= \quad 1$

and alfo the fum of his Risks $= \qquad\quad + \tfrac{224}{149}p$,

There remains his clear Gain $= - \tfrac{21}{149} + \tfrac{4224}{22201}p.$

From the Expectations of $A = \quad 4 \times \tfrac{25}{149} + \tfrac{23547}{22201}p$,

Subtracting his own Stake $= \quad 1$

and alfo the Sum of his Risks $= \qquad\qquad \tfrac{175}{194}p$,

Laftly, the Fine due to the Stock $\Big\} =$
by the lofs of the firft Game. $\qquad\qquad\qquad\qquad p$,

There remains his clear Gain $= - \tfrac{49}{149} - \tfrac{14729}{22201}p.$

The foregoing Calculation being made upon the supposition of *B* beating *A* in the beginning of the Play, which supposition neither affects *C* nor *D*, it follows that the sum of
the Gains between *B* and *A* ought to be divided equally;
and their several Gains will stand as follows,

$$\text{Gain of} \begin{cases} A = & \frac{11}{149} - \frac{2700}{22201}p, \\ B = & \frac{11}{149} - \frac{2700}{22201}p, \\ C = -\frac{5}{149} + \frac{1176}{22201}p, \\ D = -\frac{21}{149} + \frac{4224}{22201}p, \end{cases}$$

Sum of the Gains $= \quad 0 \qquad 0.$

If $\frac{11}{149} - \frac{2700}{22201}p$, which is the Gain of *A* or *B*, be
made $= 0$; then p will be found $= \frac{1937}{2700}$: From which
it follows, that if each Man's Stake be to the Fine in the
proportion of 2700 to 1937, then *A* and *B* are in this case
neither winners nor losers; but *C* wins $\frac{1}{225}$, which *D* loses.

And in the like manner may be found what the proportion
between the Stake and the Fine ought to be, to make *C* or
D play without Advantage or Disadvantage; and also what
this proportion ought to be, to make them play with any
Advantage or Disadvantage given.

Corollary I. A spectator *R* might at first in consideration of the Sum $4 + 7p$ paid him in hand, undertake to
furnish the four Gamesters with Stakes, and to pay all their
Fines.

Corollary II. If the proportion of Skill between the Gamesters be given, then their Gain or Loss may be determined by the methods used in this and the preceding Problem.

Corollary III. If there be never so many Gamesters playing on the conditions of this Problem, and the proportion of Skill between them all be supposed equal, then the
Probabilities of winning, or of being Fined, may be determined as follows.

Let

Let $\overline{B'}$, $\overline{C'}$, $\overline{D'}$, $\overline{E'}$, $\overline{F'}$, $\overline{A'}$ denote the Probabilities which B, C, D, E, F, A have of winning the Set, or of being Fined, in any number of Games; and let the Probabilities of winning or being Fined in any number of Games less by one than the preceding, be denoted by $\overline{B''}$ $\overline{C''}$ $\overline{D''}$ $\overline{E''}$ $\overline{F''}$ $\overline{A''}$: And so on. Then I say that,

$$\overline{B'} = \tfrac{1}{2}\,\overline{A''} + \tfrac{1}{4}\,\overline{A'''} + \tfrac{1}{8}\,\overline{A''''} + \tfrac{1}{16}\,\overline{A^V}$$

$$\overline{C'} = \tfrac{1}{2}\,\overline{B''} + \tfrac{1}{4}\,\overline{F'''} + \tfrac{1}{8}\,\overline{E''''} + \tfrac{1}{16}\,\overline{D^V}$$

$$\overline{D'} = \tfrac{1}{2}\,\overline{C''} + \tfrac{1}{4}\,\overline{B'''} + \tfrac{1}{8}\,\overline{F''''} + \tfrac{1}{16}\,\overline{E^V}$$

$$\overline{E'} = \tfrac{1}{2}\,\overline{D''} + \tfrac{1}{4}\,\overline{C'''} + \tfrac{1}{8}\,\overline{B''''} + \tfrac{1}{16}\,\overline{F^V}$$

$$\overline{F'} = \tfrac{1}{2}\,\overline{E''} + \tfrac{1}{4}\,\overline{D'''} + \tfrac{1}{8}\,\overline{C''''} + \tfrac{1}{16}\,\overline{B^V}$$

$$\overline{A'} = \tfrac{1}{2}\,\overline{F''} + \tfrac{1}{4}\,\overline{E'''} + \tfrac{1}{8}\,\overline{D''''} + \tfrac{1}{16}\,\overline{C^V}$$

Corollory IV.   If the Terms A, B, C, D, E, F &c. of a Series be continually decreasing, and that the Relation which each Term of the Series has to the same number of preceding ones be constantly exprest by the same number of given Fractions $\tfrac{1}{p}$, $\tfrac{1}{q}$, $\tfrac{1}{r}$, $\tfrac{1}{s}$ &c. For Example, if E be equal to $\tfrac{1}{p}D + \tfrac{1}{q}C + \tfrac{1}{r}B$, and F be also equal to $\tfrac{1}{p}E + \tfrac{1}{q}D + \tfrac{1}{r}C$, and so on: Then I say that all the Terms *Ad infinitum* of such Series as this, may be easily summed up, by following the steps of the Analysis used in this Problem; of which several Instances will be given in the Problem relating to the duration of Play.

And if the Terms of such Series be multiplyed respectively by any Series of Terms, whose last differences are equal, then the Series resulting from this multiplication is exactly summable.

And if there be two such Series or more, and the Terms of one be respectively multiplyed by the corresponding Terms of the other, then the Series resulting from this multiplication will be exactly summable.

Lastly, If there be several Series so related to one another, that each Term in the one may have to a certain number of terms in the other certain given porportions, and that the order of these proportions be constant and uniform, then will all those Series be exactly summable.   The

UNIV. OF CALIFORNIA

The foregoing Problem having been formerly Solved by me, and Printed in the *Philosophical Transactions* N° 341. Dr. *Brook Taylor*, that Excellent Mathematician, Secretary to the *Royal Society*, and my Worthy Friend, soon after communicated to me a very Ingenious Method of his, for finding the Relations which the Probabilities of winning bear to one-another, in the case of an equality of Skill between the Gamesters. The Method is as follows.

Let *B A C D* represent the four Gamesters; let also the two first Letters represent that *B* beats *A* the first Game, and the other two the order of Play.

This being supposed, the circumstances of the Gamesters will be represented in the next Game by *BCDA* or *CBDA.*

Again, the two preceding Combinations will each of them produce two more Combinations for the Game following, so that the Combinations for that Game will be four in all, *viz. BDAC, DBAC,* and *CDAB, DCAB;* which may be fitly represented by the following Scheme.

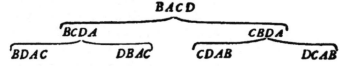

$$BACD$$
$$BCDA \qquad CBDA$$
$$BDAC \qquad DBAC \qquad CDAB \qquad DCAB$$

It appears from this Scheme, that if the Combination *CBDA* Happens, which must be in the second Game, then *B* will be in the same Circumstance wherein *A* was the Game before; the conformity of which Circumstances lies in this, that *B* is beat by one who was just come into Play when he engaged him. It appears likewise, that if the Combination *DBAC* Happens, which must be in the third Game, then *B* is again in the same Circumstance wherein *A* was two Games before.

But the Probability of the first Circumstance is $\frac{1}{2}$, and the Probability of the second is $\frac{1}{4}$.

Wherefore the Probability which *B* has of winning the Set, in any number of Games taken from the beginning, is $\frac{1}{2}$ of the Probability which *A* has of winning it in the same number of Games wanting one, taken from the beginning; as

D d   also

alſo $\frac{1}{4}$ of the Probability which he has of winning in the ſame number of Games wanting two. From which it follows, that if the Probability which *B* has of winning the Set, in Five Games for inſtance, and the Probabilities which *A* has of winning it in Four and Three, be reſpectively denoted by B^v, A'''', A''', we ſhall have the Equation, $B^v = \frac{1}{2} A'''' + \frac{1}{4} A'''$, which is conformable to what we had found before. And from the Inſpection of the ſame Scheme may likewiſe be deduced the Relations of the Probabilities of winning, as they lye between the other Gameſters. And other Schemes of this nature for any number of Gameſters may eaſily be made in imitation of this, by which the Probabilities of winning or being Fined may be determined by bare Inſpection.

PROBLEM XXXIII.

TWO Gameſters A and B, whoſe proportion of Skill is as a to b, each having a certain number of Pieces, play together on condition that as often as A Wins a Game, B ſhall give him one Piece, and that as often as B Wins a Game, A ſhall give him one Piece; and that they ceaſe not to Play till ſuch time as either one or the other has got all the Pieces of his adverſary. Now let us ſuppoſe two Spectators R and S to lay a Wager about the Ending of the Play, the firſt of them laying that the Play will be Ended in a certain number of Games which he aſſigns; the other laying to the contrary. What is the Probability that S has of Winning his Wager?

SOLUTION.

CASE I.

LET Two be the number of Pieces which each Gameſter has, let alſo Two be the number of Games about which the Wager is laid: Now becauſe two is the number of Games contended for, let $a + b$ be raiſed to its Square, *viz.* $aa + 2ab + bb$; and it is plain that the Term $2ab$ favours *S*, and that the other two are againſt him, and conſequently that the Probability he has of Winning is $\frac{2ab}{a+b|^2}$.

CASE II.

CASE II.

LET Two be the number of Pieces of each Gamester, but let Three be the number of Games upon which the Wager is laid: Then $a+b$ being raised to its Cube *viz.* $a^3 + 3aab + 3abb + b^3$, it is plain that the two Terms a^3 and b^3 are contrary to S, since they denote the number of Chances for winning three times together; 'tis plain also that the other Terms $3aab + 3abb$ are partly for him, partly against him. Let these Terms therefore be divided into their proper parts, *viz.* $3aab$ into aab, aba, baa, and $3abb$, into abb, bab, bba. Now out of these Six parts there are four which are favourable to S, *viz.* aba, baa, abb, bab or $2aab + 2abb$; from whence it follows that the Probability which S has of winning his Wager will be $\frac{2aab + 2abb}{a+b^3}$: Or dividing both Numerator and Denominator by $a+b$, it will be found to be $\frac{2ab}{a+b^2}$, which is the same as before.

CASE III.

LET Two be the number of Pieces of each Gamester, and Four the number of Games upon which the Wager is laid: Let therefore $a+b$ be raised to the fourth Power, which is $a^4 + 4a^3b + 6aabb + 4ab^3 + b^4$. The Terms $a^4 + 4a^3b + 4ab^3 + b^4$ are wholly against S, and the only Term $6aabb$ is partly for him, partly against him: Let this Term therefore be divided into its Parts, *viz.* $aabb, abab, abba, baab, baba, bbaa$; and Four of these Parts $abab, abba, baab, baba$, or $4aabb$ will be found to favours S; from which it follows, that his Probability of winning will be $\frac{4aabb}{a+b^4}$.

CASE IV.

IF Two be the number of Peices of each Gamester, and Five the number of Games about which the Wager is laid; the Probability which S has of Winning his Wager will be found to be the same as in the preceding Case, *viz.* $\frac{4aabb}{a+b^4}$.

GENE,

GENERALLY.

Let Two be the number of Pieces of each Gamester, and $2 + d$ the number of Games about which R and S contend, and it will be found that the Probability which S has of Winning will be $\dfrac{\overline{2ab}^{1+\frac{1}{2}d}}{\overline{a+b}^{2+d}}$. But if d be an odd Number, substitute $d-1$ in the room of it.

CASE V.

LET Three be the number of Pieces of each Gamester, and $3 + d$ the number of Games upon which the Wager is laid; and the Probability which S has of Winning will be $\dfrac{\overline{3ab}^{1+\frac{1}{2}d}}{\overline{a+b}^{2+d}}$. But if d be an Odd Number, you are to substitute $d-1$ in the room of it.

CASE VI.

IF the number of Pieces of each Gamester be more than Three, the Expectation of S, or the Probability there is that the Play will not be Ended in a given number of Games, may be determined in the following manner.

A General R U L E for Determining what Probability there is that the Play will not be Ended in a given number of Games.

LET n be the number of Pieces of each Gamester; let also $n + d$ be the number of Games given. Raise $a+b$ to the Power n, then cut off the two extream Terms, and multiply the remainder by $aa + 2ab + bb$: then cut off again the two Extreams, and multiply again the remainder by $aa + 2ab + bb$, still rejecting the two Extreams, and so on, making as many Multiplycations as there are Units in $\frac{1}{2}d$; Let the last Product be the Numerator of a Fraction whose Denominator is $\overline{a+b}^{n+d}$, and that Fraction will expreſs the Proba-

Probability required, or the Expectation of *S*. Still obferving that if *d* be an Odd Number, you write *d*—1 in the room of it.

EXAMPLE I.

LET Four be the number of Pieces of each Gamefter, and Ten the number of Games given: In this Cafe $n = 4$, and $n + d = 10$. Wherefore $d = 6$, and $\frac{1}{2}d = 3$. Let therefore $a + b$ be raifed to the Fourth Power, and rejecting continually the Extreams, let three Multiplications be made by $aa + 2ab + bb$. thus,

$$a^4) + 4a^3b + 6aabb + 4ab^3 (+ b^4$$
$$aa + 2ab + bb$$

$$4a^5b) + 6a^4bb + 4a^3b^3$$
$$+ 8a^4bb + 12a^3b^3 + 8aab^4$$
$$+ 4a^3b^3 + 6aab^4 (+ 4ab^5$$

$$14a^4bb + 20a^3b^3 + 14aab^4$$
$$aa + 2ab + bb$$

$$14a^6b^5) + 20a^5b^3 + 14a^4b^4$$
$$+ 28a^5b^3 + 40a^4b^4 + 28a^3b^5$$
$$+ 14a^4b^4 + 20a^3b^5 (+ 14aab^6$$

$$48a^5b^3 + 68a^4b^4 + 48a^3b^5$$
$$aa + 2ab + bb$$

$$48a^7b^3) + 68a^6b^4 + 48a^5b^5$$
$$+ 96a^6b^4 + 136a^5b^5 + 96a^4b^6$$
$$+ 48a^5b^5 + 68a^4b^6 (+ 48a^3b^7$$

$$164a^6b^4 + 232a^5b^5 + 164a^4b^6.$$

Wherefore the Probability that the Play will not be ended in Ten Games will be $\frac{164a^6b^4 + 232a^5b^5 + 164a^4b^6}{\overline{a+b}^{10}}$, which expreffion will be reduced to $\frac{560}{1024}$ or $\frac{35}{64}$, if there be an equality of Skill between the Gamefters. Now this Fraction being fubtracted from Unity, the remainder will be $\frac{29}{64}$, which will exprefs the Probability of the Play Ending in

Ten

Ten Games: And confequently it is 35 to 29, that two e-
qual Gamefters playing together, there will not be Four Stakes
loft on either fide in Ten Games.

N. B. The foregoing Operation may be very much contracted by omitting the Letters a and b, and reftoring them
after the laft Multiplication; which may be done in this manner. Make $n + \frac{1}{2}d - 1 = p$, and $\frac{1}{2}d + 1 = q$: Then an-
nex to the refpective Terms refulting from the laft Multiplication the literal Products $a^p b^q$, $a^{p-1} b^{q+1}$, $a^{p-2} b^{q+2}$ &c.
Thus in the foregoing Example, inftead of the firft Multi-
plicand $4 a^3 b + 6 aabb + 4 ab^3$, we might have taken only
$4 + 6 + 4$, and inftead of Multiplying Three times by
$aa + 2ab + bb$, we might have Multiplyed only by $1 + 2 + 1$,
which would have made the laft Terms to have been $164 +$
$232 + 164$. Now fince that n is $= 4$ and $d = 6$, p will
be $= 6$, and $q = 4$; and confequently the literal Products
to be annext to the Terms $164 + 232 + 164$ will be refpectively $a^6 b^4$, $a^5 b^5$, $a^4 b^6$, which will make the Terms refulting from the laft Multiplication to be $164 a^6 b^4 + 232 a^5 b^5$
$+ 164 a^4 b^6$, as they had been found before.

EXAMPLE II.

LET Five be the number of Pieces of each Gamefter, and
Ten the number of Games given. Let alfo the proportion of Skill between A and B be as Two to One.

Since n is $= 5$, and $n + d = 10$, it follows that d is $= 5$.
Now d being an odd number muft be-leffened by Unity,
and fuppofed $= 4$, fo that $\frac{1}{2}d = 2$. Let therefore $a + b$ be
raifed to the fifth Power; and always rejecting the extreams,
Multiply twice by $aa + 2ab + bb$, or rather by $1 + 2 + 1$;
thus,

$1) + 5 + 10 + 10 + 5 (+1$	$20 + 35 + 35 + 20$
$1 + 2 + 1$	$1 + 2 + 1$

$5) + 10 + 10 + 5$	$20) + 35 + 35 + 20$
$+ 10 + 20 + 20 + 10$	$40 + 70 + 70 + 40$
$+ 5 + 10 + 10 (+ 5$	$20 + 35 + 35 (+ 20$

| $20 + 35 + 35 + 20$ | $75 + 125 + 125 + 75$ |

Now

Now to supply the literal Products that are wanting, let $n + \frac{1}{2}d - 1$ be made $= p$, and $\frac{1}{2}d + 1 = q$, then p will be $= 6$ and $q = 3$. Wherefore the Products to be annext, *viz.* $a^p b^q$, $a^{p-1} b^{q+1}$ &c. will become $a^6 b^3$, $a^5 b^4$, $a^4 b^5$, $a^3 b^6$; and consequently the Expectation of S will be found to be

$$\frac{75 a^6 b^3 + 125 a^5 b^4 + 125 a^4 b^5 + 75 a^3 b^6}{\overline{a+b}^9}.$$

N. B. When n is an odd number, as it is in this Case, the Expectation of S will always be divisible by $a+b$. Wherefore dividing both Numerator and Denominator by $a+b$, the foregoing Expression will be reduced to

$$\frac{75 a^5 b^3 + 50 a^4 b^4 + 75 a^3 b^5}{\overline{a+b}^8}, \quad \text{or} \quad 25 a^3 b^3 \times \frac{3 aa + 2 ab + 3 bb}{\overline{a+b}^8}$$

Let now a be interpreted by 2 and b by 1, and the Expectation of S will become $\frac{3800}{6561}$.

PROBLEM XXXIV.

THE same Things being given as in the preceding Problem; to find the Expectation of R, or otherwise what the Probability is that the Play will be Ended in a given number of Games.

SOLUTION.

First, It is plain that if the Expectation of S, obtained by the preceding Problem, be subtracted from Unity, there will remain the Expectation of R.

Secondly, Since the Expectation of S decreases continually as the number of Games increases, and that the Terms we rejected in the former Problem being divided by $aa + 2ab + bb$ are the Decrement of his Expectation; it follows, that if those rejected Terms be divided continually by $\overline{a+b}^2$ they will be the Increment of the Expectation of R. Wherefore the Expectation of R may be expressed by means of those rejected Terms. Thus, in the second Example of the preceding Problem, the Expectation of R expressed by means of the rejected Terms will be found to be

$$a^3 + b^3.$$

$$\frac{a^5+b^5}{\overline{a+b}|^5} + \frac{5\,a^4b+5ab^4}{\overline{a+b}|^7} + \frac{20\,a^3bb+20\,aab^3}{\overline{a+b}|^9} \quad \text{or}$$

$$\frac{a^5+b^5}{\overline{a+b}|^5} \times \overline{1 + \frac{5\,ab}{\overline{a+b}|^2} + \frac{20\,aabb}{\overline{a+b}|^4}}.$$

In the like manner, if Six were the number of the Pieces of each Gamester, and the number of Games were Fourteen; it would be found that the Expectation of R would be

$$\frac{a^6+b^6}{\overline{a+b}|^6} \times \overline{1 + \frac{6\,ab}{\overline{a+b}|^2} + \frac{27\,aabb}{\overline{a+b}|^4} + \frac{110\,a^3b^3}{\overline{a+b}|^6} + \frac{429\,a^4b^4}{\overline{a+b}|^8}}:$$

And if Seven were the number of the Pieces of each Gamester, and the number of Games given were Fifteen; then the Expectation of R would be found to be

$$\frac{a^7+b^7}{\overline{a+b}|^7} \times \overline{1 + \frac{7\,ab}{\overline{a+b}|^2} + \frac{35\,aabb}{\overline{a+b}|^4} + \frac{154\,a^3b^3}{\overline{a+b}|^6} + \frac{637\,a^4b^4}{\overline{a+b}|^8}}.$$

N. B. The number of Terms of these Series will always be equal to $\frac{1}{2}d + 1$, if d be an even number, or to $\frac{d+1}{2}$ if it be odd.

Thirdly, All the Terms of these Series have to one another certain Relations; which being once discovered, each Term of any Series resulting from any Case of this Problem, may be easily generated from the preceding ones.

Thus in the first of the two last foregoing Series, the Numerical Coefficient belonging to the Numerator of each Term, may be derived from the preceding ones, in the following manner. Let K, L, M be the Three last Coefficients, and let N be the Coefficient of the next Term wanted; then it will be found that N in that Series will constantly be equal to $6M - 9L + 2K$. Wherefore if the Term which would follow $\frac{429\,a^4b^4}{\overline{a+b}|^8}$, in the Case of Sixteen Games given, were desired; then make $M = 429$, $L = 110$, $K = 27$, and the following Coefficient will be found 1638. From whence it appears that the Term it self would be $\frac{1638\,a^5b^5}{\overline{a+b}|^{10}}$.

Likewise, in the second of the two foregoing Series, if the Law by which each Term is related to the preceding ones were

were demanded, it might be thus found. Let K, L, M, be the Coefficients of the three laſt Terms, and N the Coefficient of the Term deſired; then N will in that Series, conſtantly be equal to $7M - 14L + 7K$, or to $\overline{M - 2L + K} \times 7$. Now this Coefficient being obtained, the Term to which it belongs is formed immediately.

But if the general Law, by which each Coefficient is generated from the preceding ones, be demanded, it will be expreſt as follows. Let n be the number of Pieces of each Gameſter: Then each Coefficient contains

n times the laſt,

$- n \times \frac{n-1}{2}$ times the laſt but one,

$+ n \times \frac{n-4}{2} \times \frac{n-1}{3}$ times the laſt but two,

$- n \times \frac{n-5}{2} \times \frac{n-6}{3} \times \frac{n-7}{4}$ times the laſt but three;

$+ n \times \frac{n-6}{2} \times \frac{n-7}{3} \times \frac{n-8}{4} \times \frac{n-9}{5}$ times the laſt but four,

&c.

Thus the number of Pieces of each Gameſter being Six, the firſt Term n would be $= 6$, the ſecond Term $n \times \frac{n-1}{2}$ would be $= 9$, the third Term $n \times \frac{n-4}{2} \times \frac{n-5}{3}$ would be $= 2$; the reſt of the Terms vaniſhing in this Caſe. Wherefore if K, L, M are the three laſt Coefficients, the Coefficient of the following Term will be $6M - 9L + 2K$.

Fourthly, The Coefficient of any Term of theſe Series may be found, independently from any relation they may have to the preceding ones: In order to which it is to be obſerved that each Term of theſe Series is proportional to the Probability of the Plays Ending in a certain number of Games precifely: Thus in the Series which expreſſes the Expectation of R, when each Gameſter is ſuppoſed to have Six Pieces, *viz.*

$$\frac{a^6 + b^6}{a + b^6} \times 1 + \frac{6ab}{a + b^6} + \frac{27aab b}{a + b^6} + \frac{110a^3b^3}{a + b^6} + \frac{429a^4b^4}{a + b^8},$$

the laſt Term, being multiplied by the common Multiplicator $\frac{a^6 + b^6}{a + b^6}$ fet down before the Series, that is the Product $\frac{429a^4b^4 \times \overline{a^6 + b^6}}{a + b^{10}}$, denotes the Probability of the Plays

F f   End-

Ending in Fourteen Games precisely. Wherefore if that Term were desired which expresses the Probability of the Plays Ending in Twenty Games precisely, or in any number of Games denoted by $n+d$, I say that the Coefficient of that Term will be,

$$\frac{1}{1} \times \frac{n}{1} \times \frac{n+d-1}{2} \times \frac{n+d-2}{3} \times \frac{n+d-3}{4} \ \&c. \ \text{continued}$$

to so many Terms as there are Units in $\frac{1}{2}d + 1$.

$$- \frac{1}{1} \times \frac{3n}{1} \times \frac{n+d-1}{2} \times \frac{n+d-2}{3} \times \frac{n+d-3}{4} \ \&c. \ \text{continued}$$

to so many Terms as there are Units in $\frac{1}{2}d + 1 - n$.

$$+ \frac{1}{1} \times \frac{5n}{1} \times \frac{n+d-1}{2} \times \frac{n+d-2}{3} \times \frac{n+d-3}{4} \ \&c. \ \text{continued}$$

to so many Terms as there are Units in $\frac{1}{2}d + 1 - 2n$.

$$- \frac{1}{1} \times \frac{7n}{1} \times \frac{n+d-1}{2} \times \frac{n+d-2}{3} \times \frac{n+d-3}{4} \ \&c. \ \text{continued}$$

to so many Terms as there are Units in $\frac{1}{2}d + 1 - 3n$. &c.

Let now $n+d$ be supposed $= 20$, n being already supposed $= 6$, then the Coefficient demanded will be found from the general Rule to be,

$$\frac{1}{1} \times \frac{6}{1} \times \frac{19}{2} \times \frac{18}{3} \times \frac{17}{4} \times \frac{16}{5} \times \frac{15}{6} \times \frac{14}{7} = 23256$$
$$- \frac{1}{1} \times \frac{18}{1} \qquad\qquad\qquad\qquad\qquad = 18$$

Wherefore the Coefficient demanded will be $23256 - 18 = 23238$: And then the Term it self to which this Coefficient does belong, will be $\frac{23238\, a^7 b^7}{a + b|^{14}}$. Consequently the Probability of the Plays Ending in Twenty Games precisely will be $\frac{a^6 + b^6}{a+b|^6} \times \frac{23238\, a^7 b^7}{a+b|^{14}}$.

Fifthly, By the help of the two Methods explained in this Problem (whereof the first is for finding the relation which any Term of the Series resulting from the Problem, has to a certain number of preceding ones; and the second for finding any Term of the Series independently from any other Term) together with the Method of summing up any given

num-

number of Terms of thefe Series, (which fhall be explained
in its place); the Probability of the Plays Ending in any given number of Games, will be found much more readily than
can be done by either of the two firft Methods taken
fingly.

PROBLEM XXXV.

SUppofing A *and* B *to play together, till fuch time as Four
Stakes are Won or Loft on either fide: What muft be their proportion of Skill, to make it as Probable that the Play will be Ended in Four Games, as not?*

SOLUTION.

THE Probability of the Play Ending in Four Games, is
by the preceding Problem $\frac{a^4+b^4}{a+b|^4}$; Now becaufe, by
Hypothefis, it is to be an equal Chance whether the Play
Ends or Ends not in Four Games; let this expreffion of the
Probability be made equal to $\frac{1}{2}$; And we fhall have this

Equation $\frac{a^4+b^4}{a+b|^4} = \frac{1}{2}$, which, making $b, a :: 1, z$, is re-
duced to $\frac{z^4+1}{z+1|^4} = \frac{1}{2}$, or $z^4 - 4z^3 - 6zz - 4z + 1 = 0$.
Let $12zz$ be added on both fides the Equation, then will
$z^4 - 4z^3 + 6zz - 4z + 1$ be $= 12zz$; and extracting
the fquare Root on both fides, it will be reduced to this
Quadratick Equation $zz - 2z + 1 = z\sqrt{12}$, whofe double
Root is $z = 5,274$ and $\frac{1}{5,274}$. Wherefore whether the Skill
of *A* be to that of *B* as 5.274 to 1, or as 1 to 5.274, there
will be an equality of Chance for the Play to be Ended or
not Ended in Four Games.

PROBLEM XXXVI.

SUppofing *that* A *and* B *Play till fuch time as Four Stakes
are Won or Loft: What muft be their proportion of Skill,
to make it a Wager of Three to One, that the Play will be Ended in Four Games?*

SOLU-

SOLUTION.

THE Probability of the Plays Ending in Four Games, arifing from the number of Games Four, from the number of Stakes Four, and from the proportion of Skill is $\frac{a^4+b^4}{a+b|^4}$. The fame Probability arifing from the odds of Three to One, is $\frac{3}{4}$. Wherefore $\frac{a^4+b^4}{a+b|^4} = \frac{3}{4}$, and fuppofing $b; a :: 1, z$, the foregoing Equation will be changed into $\frac{z^4+1}{z+1|^4} = \frac{3}{4}$, or $z^4 - 12\,z^3 - 18\,zz - 12\,z + 1 = 0$. Let $56\,zz$ be added on both fides the Equation, then we fhall have $z^4 - 12\,z^3 + 38\,zz - 12\,z + 1 = 56\,zz$. And Extracting the fquare Root on both fides, we fhall have $zz - 6z + 1 = z\sqrt{56}$, the Roots of which Equation will be found 13.407 and $\frac{1}{13.407}$. Wherefore, whether the Skill of A be to that of B as 13.407 to 1, or as 1 to 13.407, 'tis a Wager of Three to One, that the Play will be ended in Four Games.

PROBLEM XXXVII.

Suppofing that A and B Play till fuch time as Four Stakes are Won or Loft; What muft be their proportion of Skill, to make it an equal Wager that the Play will be Ended in Six Games?

SOLUTION.

THE Probability of the Plays Ending in Six Games, arifing from the given number Six, from the number of Stakes Four, and from the proportion of Skill, is $\overline{\frac{a^4+b^4}{a+b|^4} \times 1 + \frac{4ab}{a+b|^2}}$. The fame Probability arifing from an equality of Chance for Ending or not Ending in Six Games, is equal to $\frac{1}{2}$, from whence refults the Equation $\overline{\frac{a^4+b^4}{a+b|^4} \times 1 + \frac{4ab}{a+b|^2}} = \frac{1}{2}$ which by making $b, a :: 1, z$ may be changed into the following, viz. $z^6 + 6z^5 - 13z^4 - 20z^3 - 13zz + 6z + 1 = 0$.

In

In this Equation, the Coefficients of the Terms equally diſtant from the Extreams being the ſame, let it be ſuppoſed that the Equation is generated from the Multiplication of two other Equations of the ſame nature, *viz.* $zz - yz + 1 = 0$, and $z^4 + pz^3 + qzz + pz + 1 = 0$. Now the Equation reſulting from the Multiplication of theſe two will be

$$z^6 \quad \begin{matrix} -y \\ +p \end{matrix} z^5 \begin{matrix} +1 \\ -pyz^4 \\ +q \end{matrix} + \begin{matrix} 2p \\ z^3 -pyzz \\ -qy \end{matrix} \begin{matrix} +1 \\ +q \end{matrix} z^1 \begin{matrix} +p \\ -pyzz \\ +q \end{matrix} \begin{matrix} +p \\ z+1=0, \\ -y \end{matrix}$$

which being compared with the firſt Equation, we ſhall have
$p - y = 6$, $1 - py + q = -13$, $2p - qy = -20$.

From hence will be deduced a new Equation, *viz.* $y^3 + 6yy - 16y - 32 = 0$, one of whoſe Roots will be 2.9644; which being ſubſtituted in the Equation $zz - yz + 1 = 0$, we ſhall at laſt come to the Equation $zz - 2.9644\,z + 1 = 0$, of which the two Roots will be 2.576 and $\frac{1}{2.576}$. It follows therefore, that if the Skill of either Gameſter be to that of the other as 2.576 to 1; there will be an equal Chance for Four Stakes to be Loſt, or not to be Loſt, in Six Games.

Corollary. If the Coefficients of the Extream Terms of an Equation, and likewiſe the Coefficients of the other Terms equally diſtant from the Extreams, be the ſame, that Equation will be reducible to another, in which the Dimenſions of the higheſt Term will not exceed half the Dimenſions of the higheſt Term in the former.

PROBLEM XXXVIII.

S*Uppoſing* A *and* B, *whoſe proportion of Skill is as* a *to* b, *to Play together till ſuch time as* A *either Wins a certain number* q *of Stakes, or* B *ſome other number* p *of them: What is the Probability that the Play will not be Ended in a given number of Games?*

SOLUTION.

T AKE the Binomial $a+b$, and rejecting continually thoſe Terms in which the Dimenſions of the quantity a exceed the Dimenſions of the quantity b by q, rejecting alſo

G g

thoſe

thofe Terms in which the Dimenfions of the quantity b exceed the Dimenfions of the quantity a by p; multiply conftantly the remainder by $a+b$, and make as many Multiplications, as there are Units in the given number of Games wanting one. Then fhall the laft Product be the Numerator of a Fraction expreffing the Probability required; the Denominator of which Fraction always being the Binomial $a+b$ raifed to that Power which is denoted by the given number of Games.

EXAMPLE.

LET p be $= 3$, $q = 2$, and let the given number of Games be $= 7$. Let the following Operation be made according to the foregoing directions.

$$a + b$$
$$a + b$$

$$aa) + 2ab + bb$$
$$a + b$$

$$2aab + 3abb + (+ b^3$$
$$a + b$$

$$2a^3b) + 5aabb + 3ab^3$$
$$a + b$$

$$5a^3bb + 8aab^3 + (3ab^4$$
$$a + b$$

$$5a^4bb) + 13a^3b^3 + 8aab^4$$
$$a + b$$

$$13a^4b^3 + 21a^3b^4 (+ 8aab^5.$$

From this Operation we may conclude, that the Probability that the Play will not be Ended in Seven Games is equal to $\frac{13a^4b^3 + 21a^3b^4}{a+b^7}$. Now if an Equality of Skill be fuppofed between A and B, the Expreffion of this Probability

bility will be reduced to $\frac{33+11}{128}$ or $\frac{17}{64}$: Wherefore the Probability that the Play will End in Seven Games will be $\frac{47}{64}$; from which it follows that 'tis 47 to 17 that in Seven Games, either A wins two Stakes or B wins three.

PROBLEM XXXIX.

THE same Things being suppos'd as in the preceding Problem, to find the Probability of the Plays being Ended in a given number of Games.

SOLUTION.

First, If the Probability of the Plays not being Ended in the given number of Games be subtracted from Unity, there will remain the Probability of its being Ended in the same number of Games.

Secondly, This Probability may be expressed by means of the Terms rejected in the Operation belonging to the preceding Problem; Thus, if the number of Stakes be Three and Two, the Probability of the Plays being Ended in Seven Games may be expressed as follows.

$$\frac{aa}{\overline{a+b}|^2} \times \overline{1 + \frac{2ab}{\overline{a+b}|^2} + \frac{5aabb}{\overline{a+b}|^4}}.$$
$$+ \frac{b^3}{\overline{a+b}|^3} \times \overline{1 + \frac{3ab}{\overline{a+b}|^2} + \frac{8aabb}{\overline{a+b}|^4}}.$$

Supposing a and b both equal to Unity, the sum of the first Series will be $= \frac{29}{64}$, and the sum of the second will be $= \frac{18}{64}$; which two sums being added together, the aggregate $\frac{47}{64}$ expresses the Probability that in Seven Games either A shall win Two Stakes or B Three.

Thirdly, The Probability of the Plays being Ended in a certain number of Games, or sooner, is always composed of a double Series, when the Stakes are unequal; which double Series is reduced to a single one, in the Case of equality of Stakes.

The first Series always expresses the Probability there is that A, in a given number of Games, or sooner, may win of B the

num-

number q of Stakes, excluding the Probability there is, that B, before that time, may be in a circumstance of winning the number p of Stakes. Both which Probabilities are not inconsistent together; for A in Fifteen Games, for Instance, or sooner, may win Two Stakes of B, though B before that time may have been in a circumstance of winning Three Stakes of A.

The second Series always expresses the Probability there is that B, in that given number of Games or sooner, may win of A a certain number p of Stakes, excluding the Probability there is that A, before that time, may win of B the number q of Stakes.

The first Terms of each Series may be represented respectively by the following Terms.

$$\frac{a^q}{\overline{a+b}|^q} \times \overline{1 + \frac{q\,a\,b}{a+b|^2} + \frac{q\times q + 3\times a\,a\,b\,b}{1\times1\times a + b|^4} + \frac{q\times q + 4\times q + 5\times a^3b^3}{1\times2\times3\times a + b|^6}} \ \&c. \ \text{and}$$

$$\frac{b^p}{\overline{a+b}|^p} \times \overline{1 + \frac{p\,a\,b}{a+b|^2} + \frac{p\times p + 3\times a\,a\,b\,b}{1\times2\times a + b|^4} + \frac{p\times p + 4\times p + 5\times a^3b^3}{1\times2\times3\times a + b|^6}} \ \&c.$$

Each of these Series continuing in that regularity, till such time as there be a number p of Terms taken in the first, and a number q of Terms taken in the second; after which the Law of the continuation breaks off.

Now in order to find any of the Terms following in either of these Series, proceed thus; let $p + q - 2$ be called l; let the Coefficient of the Term desired be T; let also the Coefficients of the preceding Terms taken in an inverted order be S, R, Q, P &c. Then will T be equal to $l\,S - \frac{l-1}{1} \times \frac{l-2}{2} R + \frac{l-2}{1} \times \frac{l-3}{2} \times \frac{l-4}{3} Q - \frac{l-3}{1} \times \frac{l-4}{2} \times \frac{l-5}{3} \times \frac{l-6}{4} P$ &c. Thus if p be $= 3$, and $q = 2$, then l will be $3 + 2 - 2 = 3$. Wherefore $l\,S - \frac{l-1}{1} \times \frac{l-2}{2} \times R$ would in this Case be equal to $3\,S - R$; which shews that the Coefficient of any Term desired would be constantly three times the last, *minus* once the last but one.

To apply this, let it be required to find what Probability there is, that in Fifteen Games or sooner, either A shall win two Stakes of B, or B three Stakes of A; or which is all one, to find what Probability there is, that the Play shall end in Fifteen Games or sooner, A and B resolving to Play, till such time as A either wins three Stakes, or B two.

Let

Let Two and Three, in the two foregoing Series, be fub-flituted refpectively in the room of q and p; then the three firft Terms of the firft Series will be, fetting afide the common Multiplicator, $1 + \frac{2ab}{a+b|^2} + \frac{5aabb}{a+b|^4}$: Likewife the two

firft Terms of the fecond will be $1 + \frac{3ab}{a+b|^2}$. Now becaufe the Coefficient of any Term defired in each Series, is refpectively three times the laft, *minus* once the laft but one, it follows, that the next Coefficient in the firft Series will be 13, and by the fame rule the next to it 34, and fo on. In the fame manner the next Coefficient in the fecond Series will be found to be 8, and the next to it 21, and fo on. Wherefore, reftoring the common Multiplicators, the two Series will be

$$\frac{a^2}{a+b|^2} \times 1 + \frac{2ab}{a+b|^2} + \frac{5aabb}{a+b|^4} + \frac{13a^3b^3}{a+b|^6} + \frac{34a^4b^4}{a+b|^8} + \frac{89a^5b^5}{a+b|^{10}}$$
$$+ \frac{233a^6b^6}{a+b|^{12}}.$$

$$\frac{b^3}{a+b|^3} \times 1 + \frac{3ab}{a+b|^2} + \frac{8aabb}{a+b|^4} + \frac{21a^3b^3}{a+b|^6} + \frac{55a^4b^4}{a+b|^8} + \frac{144a^5b^5}{a+b|^{10}}$$
$$+ \frac{377a^6b^6}{a+b|^{12}}.$$

If we fuppofe an equality of Skill between A and B, the fum of the firft Series will be $\frac{18778}{32768}$, the fum of the fecond will be $\frac{12393}{32768}$, and the aggregate of thefe two fums will be $\frac{31171}{32768}$, which will exprefs the Probability of the Plays Ending in Fifteen Games or fooner. This laft Fraction being fubtracted from Unity, there will remain $\frac{1597}{32768}$, which expreffes the Probability of the Plays continuing for Fifteen Games at leaft: Wherefore 'tis 31171 to 1597, or 39 to 2 nearly, that one of the two equal Gamefters, that fhall be pitch'd upon, fhall in Fifteen Games, or fooner, either win Two Stakes of his adverfary, or lofe Three to him.

N. B. The Index of the Denominator in the laft Term of each Series, and the Index of the common Multiplicator pre-

H h   fixt

fixt to it, being added together, muſt either equal the number of Games given, or be leſs than it by Unity. Thus, in the firſt Series, the Index 12 of the Denominator of the laſt Term, and the Index 2 of the common Multiplicator being added together, the ſum is 14, which is leſs by Unity than the number of Games given. So likewiſe in the ſecond Series, the Index 12 of the Denominator of the laſt Term, and the Index 3 of the common Multiplicator being added together, the ſum is 15, which preciſely equals the number of Games given.

It is carefully to be obſerved, that theſe two Series taken together, expreſs the Expectation of one and the ſame Perſon, and not of two different Perſons; that is properly the Expectation of a ſpectator who lays a Wager that the Play will be Ended in a given number of Games. Yet in one Caſe they may expreſs the Expectations of two different Perſons: For Inſtance, of the Gameſters themſelves, provided that both Series be continued infinitely; for in that Caſe, the firſt Series infinitely continued will expreſs the Probability that the Gameſter *A* may ſooner win two Stakes of *B*, than that he may loſe three to him: Likewiſe the ſecond Series infinitely continued will expreſs the Probability that the Gameſter *B* may ſooner win three Stakes of *A*, than that he may loſe two to him. And it will be found, when we come to treat of the method of ſumming up theſe Series, that the firſt Series infinitely continued will be to the ſecond infinitely continued, in the proportion of $aa \times \overline{aa+ab+bb}$ to $b^3 \times \overline{a+b}$; that is, in the Caſe of an equality of Skill, as three to two; which is conformable to what we have ſaid in our IX*th*. Problem.

Fourthly, Any Term of theſe Series may be found independently from any of the preceding ones: For if a Wager be laid that *A* ſhall either win a certain number of Stakes denominated by *q*, or that *B* ſhall win a certain number of them denominated by *p*, and that the number of Games given be expreſſed by *q*+*d*; then I ſay that the Coefficient of any Term in the fiſt Series, anſwering to that number of Games, will be

$$+ \; \frac{1}{1} \times \frac{q}{1} \times \frac{q+d-1}{2} \times \frac{q+d-2}{3} \times \frac{q+d-3}{4} \; \&c.$$ continued to ſo many Terms as there are Units in $\frac{1}{2}d + 1$.

$$-\frac{1}{1}\times\frac{q+2p}{1}\times\frac{q+d-1}{2}\times\frac{q+d-2}{3}\times\frac{q+d-3}{4}\text{ &c. conti-}$$

nued to fo many Terms as there are Units in $\frac{1}{2}d+1-p$.

$$+\frac{1}{1}\times\frac{3q+2p}{1}\times\frac{q+d-1}{2}\times\frac{q+d-2}{3}\times\frac{q+d-3}{4}\text{ &c. continu-}$$

ed to fo many Terms as there are Units in $\frac{1}{2}d+1-p-q$:

$$-\frac{1}{1}\times\frac{3q+4p}{1}\times\frac{q+d-1}{2}\times\frac{q+d-2}{3}\times\frac{q+d-3}{4}\text{ &c. conti-}$$

nued to fo many Terms as there are Units in $\frac{1}{2}d+1-2p-q$.

$$+\frac{1}{1}\times\frac{5q+4p}{1}\times\frac{q+d-1}{2}\times\frac{q+d-2}{3}\times\frac{q+d-3}{4}\text{ &c. continued}$$

to fo many Terms as there are Units in $\frac{1}{2}d+1-2p-2q$.

$$-\frac{1}{1}\times\frac{5q+6p}{1}\times\frac{q+d-1}{2}\times\frac{q+d-2}{3}\times\frac{q+d-3}{4}\text{ &c. continu-}$$

ed to fo many Terms as there are Units in $\frac{1}{2}d+1-3p-2q$.

$$+\frac{1}{1}\times\frac{7q+6p}{1}\times\frac{q+d-1}{2}\times\frac{q+d-2}{3}\times\frac{q+d-3}{4}\text{ &c. conti-}$$

nued to fo many Terms as there are Units in $\frac{1}{2}d+1-3p-3q$.
&c.

And the fame Law will hold for the other Series, calling $p+s$ the number of Games given, and changing q into p, and p into q, as alfo d into s.

But when it Happens that d is an odd number, fubftitute $d-1$ in the room of it, and the like for s.

PROBLEM XL.

IF A and B, *whofe proportion of Skill is fuppofed as* a *to* b, *play together: What is the Probability that one of them, fuppofe* A, *may in a number of Games not exceeding a number given, win of* B *a certain number of Stakes? Leaving it wholly indifferent whether* B *before the expiration of thofe Games may or may not have been in a circumftance of winning the fame, or any other number of Stakes of* A.

SOLUTION.

SUppofing n to be the number of Stakes which A is to win of B, and $n+d$ the given number of Games; let $a+b$ be raifed to the Power whofe Index is $n+d$: Then if d be.

an.

an odd Number, take fo many Terms of that Power as there are Units in $\frac{d+1}{2}$; take alfo as many of the Terms next following as have been taken already, but prefix to them, in an inverted order, the Coefficients of the preceding Terms. But if *d* be an even number, take fo many Terms of the faid Power as there are Units in $\frac{1}{2}d+1$; then take as many of the Terms next following as there are Units in $\frac{1}{2}d$, and prefix to them, in an inverted order, the Coefficients of the preceding Terms, omitting the laft of them; and thofe Terms taken all together will compofe the Numerator of a Fraction expreffing the Probability required, its Denominator being $\overline{a+b}|^{n+d}$.

EXAMPLE I.

SUppofing the number of Stakes which *A* is to win to be Three, and the given number of Games to be Ten; let $a+b$ be raifed to the tenth Power, *viz.* $a^{10} + 10\,a^9 b + 45\,a^8 bb + 120\,a^7 b^3 + 210\,a^6 b^4 + 252\,a^5 b^5 + 210\,a^4 b^6 + 120\,a^3 b^7 + 45\,aab^8 + 10\,ab^9 + b^{10}$. Then by reafon that *n* is $= 3$ and $n+d = 10$, it follows that *d* is $= 7$, and $\frac{d+1}{2} = 4$. Wherefore let the Four firft Terms of the faid Power be taken, *viz.* $a^{10} + 10\,a^9 b + 45\,a^8 bb + 120\,a^7 b^3$, and let the Four Terms next following be taken likewife, without regard to their Coefficients; then prefix to them, in an Inverted order, the Coefficients of the preceding Terms: Thus the Four Terms following with their new Coefficients, will be $120\,a^6 b^4 + 45\,a^5 b^5 + 10\,a^4 b^6 + 1\,a^3 b^7$. And the Probability which *A* has of winning Three Stakes of *B* in Ten Games, or fooner, will be expreffed by the following Fraction,

$$\frac{a^{10}+10a^9b+45a^8bb+120a^7b^3+120a^6b^4+45a^5b^5+10a^4b^6+a^3b^7}{\overline{a+b}|^{10}},$$

which, in the Cafe of an equality of Skill between *A* and *B*, will be reduced to $\frac{352}{1024}$ or $\frac{11}{32}$.

EXAMPLE II.

SUppofing the number of Stakes which A is to win to be Four, and the given number of Games to be Ten; let $a+b$ be raifed to the tenth Power, and by reafon that n is $= 4$, and $n+d = 10$, it follows, that $d = 6$ and $\frac{1}{2}d + 1 = 4$; wherefore let the Four firft Terms of the faid Power be taken, *viz.* $a^{10} + 10 a^9 b + 45 a^8 bb + 120 a^7 b^3$; take alfo Three of the Terms following, but prefix to them, in an invefted order, the Coefficients of the Terms already taken, omitting the laft of them. Hence the Three Terms following with their new Coefficients will be $45 a^6 b^4 + 10 a^5 b^5 + 1 a^4 b^6$. And the Probability which A has of winning Four Stakes of B, in Ten Games or fooner, will be expreffed by the following Fraction

$$\frac{a^{10} + 10a^9 b + 45 a^8 bb + 120 a^7 b^3 + 45 a^6 b^4 + 10 a^5 b^5 + 1 a^4 b^6}{\overline{a+b}^{10}}$$

which, in the Cafe of an equality of Skill between A and B, will be reduced to $\frac{232}{1024}$ or $\frac{29}{128}$.

Another SOLUTION.

SUppofing, as before, that n be the number of Stakes which A is to win, and that the given number of Games be $n+d$ the Probability which A has of winning will be expreffed by the following Series, *viz.*

$$\frac{a^n}{\overline{a+b}^n} \times 1 + \frac{n \times b}{\overline{a+b}^2} + \frac{n \times \overline{n+3} \times aa bb}{1 \times 2 \times \overline{a+b}^4} + \frac{n \times \overline{n+4} \times \overline{n+5} \times a^3 b^3}{1 \times 2 \times 3 \times \overline{a+b}^6}$$
$$+ \frac{n \times \overline{n+5} \times \overline{n+6} \times \overline{n+7} \times a^4 b^4}{1 \times 2 \times 3 \times 4 \times \overline{a+b}^8} \quad \&c.$$

which Series ought to be continued to fo many Terms as there are Units in $\frac{1}{2}d + 1$; always obferving to fubftitute $d-1$ in the room of d, in cafe d be an odd number, or which is the fame thing, taking fo many Terms as there are Units in $\frac{d+1}{2}$.

Now

Now suppofing, as in the firft Example of the preceding Solution, that Three is the number of Stakes, and Ten the given number of Games, as alfo that there is an equality of Skill between A and B, the foregoing Series will become

$$\frac{1}{8} \times \overline{1 + \frac{3}{4} + \frac{9}{16} + \frac{27}{64}} = \frac{11}{32}, \text{ as before.}$$

REMARK.

MOnfieur *de Monmort*, in the Second Edition of his Book of Chances, having given a very handfom, Solution of the Problem relating to the duration of Play, (which Solution is coincident with that of Monfieur *Nicolas Bernoully*, to be feen in that Book) and the Demonftration of it being very naturally deduced from our firft Solution of the foregoing Problem, I thought the Reader would be well pleafed to fee it transferred to this place.

Let it therefore be propofed to find the number of Chances there are, for A either to win Two Stakes of B, or for B to win Three of A in Fifteen Games.

The number of Chances required is expreffed by two Branches of Series; all the Series of the firft Branch taken together exprefs the number of Chances there are for A to win Two Stakes of B, exclufive of the number of Chances there are for B, before that time, to win Three Stakes of A. All the Series of the fecond Branch taken together exprefs the number of Chances there are for B to win Three Stakes of A, exclufive of the number of Chances there are for A, before that time, to win Two Stakes of B.

Firft Branch of SERIES.

$a^{15} \quad a^{14}b \quad a^{13}b^2 \quad a^{12}b^3 \quad a^{11}b^4 \quad a^{10}b^5 \quad a^9b^6 \quad a^8b^7 \quad a^7b^8 \quad a^6b^9 \quad a^5b^{10} \quad a^4b^{11} \quad a^3b^{12} \quad a^2b^{13}$

$1 + 15 + 105 + 455 + 1365 + 3003 + 5005 + 5005 + 3003 + 1365 + 455 + 105 + 15 + 1$
$\quad\quad -1 - 15 - 105 - 455 - 455 - 105 - 15 - 1$
$\quad\quad\quad + 1 + 15 + 15 + 1$

Second Branch of SERIES.

$b^{15} \quad b^{14}a \quad b^{13}a^2 \quad b^{12}a^3 \quad b^{11}a^4 \quad b^{10}a^5 \quad b^9a^6 \quad b^8a^7 \quad b^7a^8 \quad b^6a^9 \quad b^5a^{10} \quad b^4a^{11} \quad b^3a^{12}$

$1 + 15 + 105 + 455 + 1365 + 3003 + 5005 + 3003 + 1365 + 455 + 105 + 15 + 1$
$\quad\quad - 1 - 15 - 105 - 455 - 1365 - 455 - 105 - 15 - 1$
$\quad\quad\quad + 1 + 15 + 1$

The

The literal Quantities, which are commonly annext to the numerical ones, are here written on the top of them; which is done, to the end that each Series being contained in one line, the dependency they have upon one another, may thereby be made the more conspicuous.

The first Series of the first Branch expresses the number of Chances there are for *A* to win Two Stakes of *B*, including the number of Chances there are for *B*, before the expiration of the Fifteen Games, to be in a circumstance of winning Three Stakes of *A*; which number of Chances may be deduced from our foregoing Problem.

The second Series of the first Branch is a part of the first; and expresses the number of Chances there are, for *B* to win Three Stakes of *A*, out of the number of Chances there are for *A* in the first Series, to win Two Stakes of *B*. It is to be observed about this Series, *First*, that the Chances of *B* expressed by it are not restrained to Happen in any Order, that is, either before or after *A* has won Two Stakes of *B*. *Secondly*, that the literal Products belonging to it are the same with those of the corresponding Terms of the first Series. *Thirdly*, that it begins and ends at an interval from the first and last Terms of the first Series equal to the number of Stakes which *B* is to win. *Fourthly*, that the numbers belonging to it are the numbers of the first Series repeated in order, and continued to one half of its Terms; after which those numbers return in an inverted order to the end of that Series: Which is to be understood in case the number of its Terms should Happen to be even, for if it should Happen to be odd, then that order is to be continued to the greatest half, after which the return is made by omitting the last number. *Fifthly*, that all the numbers of it are Negative.

The Third Series of the first Branch is a part of the second, and expresses the number of Chances there are for *A* to win Two Stakes of *B*; out of the number of Chances there are in the second Series, for *B*, to win Three Stakes of *A*; with this difference, that it begins and ends at an interval from the first and last Terms of the second Series, equal to the number of Stakes which *A* is to win; and that the Terms of it are all Positive.

It ·

It is to be obferved in general that, let the number of thefe Series be what it will, the Interval between the beginning of the firft and the beginning of the fecond, is to be equal to the number of Stakes which *B* is to win; and that the Interval between the beginning of the fecond and the beginning of the third, is to be equal to the number of Stakes which *A* is to win; and that thefe Intervals recurr alternately in the fame Order. It is to be obferved likewife, that all thefe Series are alternately Pofitive and Negative.

All the Obfervations made upon the firft Branch of Series belonging alfo to the fecond, it would be needlefs to fay any more of them.

Now the fum of all the Series of the firft Branch, being added to the fum of all the Series of the fecond, the aggregate of thefe fums will be the Numerator of a Fraction expreffing the Probability of the Plays terminating in the given number of Games; of which Fraction the Denominator is the Binomial $a+b$ raifed to a Power, whofe Index is equal to that given number of Games. Thus, fuppofing that, in the Cafe of this Problem, both a and b are equal to Unity, the fum of the Series in the firft Branch will be 18778, the fum of the Series in the fecond will be 12393; and the aggregate of both 31171: And the Fifteenth Power of 2 being 32768, it follows, that the Probability of the Plays terminating in Fifteen Games will be $\frac{31171}{32768}$, which being fubtracted from Unity, the remainder will be $\frac{1597}{32768}$: From whence we may conclude, that 'tis a Wager of 31171 to 1597, that either *A* in Fifteen Games fhall win Two Stakes of *B*, or *B* win Three Stakes of *A*: Which is conformable to what we had before found in our XXXIX*th*. Problem.

PROBLEM XLI.

TO *Find what Probability there is, that in a given number of Games,* A *may be winner of a certain number* q *of Stakes; and at fome other time,* B *may likewife be winner of the number* p *of Stakes, fo that both circumftances may Happen.*

SOLU-

SOLUTION.

FIND, by our XL*th* Problem, the Probability which *A* has of winning, without any Limitation, the number *q* of Stakes: Find alſo by our XXXIV*th* Problem the Probability which *A* has of winning that number of Stakes before *B* may Happen to win the number *p*; then from the firſt Probability ſubtracting the ſecond, the remainder will expreſs the Probability there is, that both *A* and *B* may be in a circumſtance of winning, but *B* before *A*. In the like manner, from the probability which *B* has of winning without any Limitation, ſubtracting the Probability which he has of winning before *A*, the remainder will expreſs the Probability there is, that both *A* and *B* may be in a circumſtance of winning, but *A* before *B*. Wherefore adding theſe two remainders together, their ſum will expreſs the Probability required.

Thus, if it were required to find what Probability there is, that in Ten Games *A* may win Two Stakes, and that at ſome other time *B* may win Three. The firſt Series will be found to be,

$$\frac{aa}{a+b|^2} \times \overline{1 + \frac{2ab}{a+b|^2} + \frac{5aabb}{a+b|^4} + \frac{14a^3b^3}{a+b|^6} + \frac{42a^4b^4}{a+b|^8}}.$$

The ſecond Series will likewiſe be found to be

$$\frac{aa}{a+b|^2} \times \overline{1 + \frac{2ab}{a+b|^2} + \frac{5aabb}{a+b|^4} + \frac{13a^3b^3}{a+b|^6} + \frac{34a^4b^4}{a+b|^8}}.$$

The difference of theſe Series being $\frac{aa}{a+b|^2} \times \overline{\frac{a^3b^3}{a+b|^6} + \frac{8a^4b^4}{a+b|^8}}$ expreſſes the firſt part of the Probability required, which, in the Caſe of an equality of Skill between the Gameſters, would be reduced to $\frac{3}{256}$.

The Third Series is as follows,

$$\frac{b^3}{a+b|^3} \times \overline{1 + \frac{3ab}{a+b|^2} + \frac{9aabb}{a+b|^4} + \frac{28a^3b^3}{a+b|^6}}.$$

The Fourth Series is

$$\frac{b^3}{a+b|^3} \times \overline{1 + \frac{3ab}{a+b|^2} + \frac{8aabb}{a+b|^4} + \frac{21a^3b^3}{a+b|^6}}.$$

The difference of theſe two Series being $\frac{b^3}{a+b|^3} \times \overline{\frac{aabb}{a+b|^4} + \frac{7a^3b^3}{a+b|^6}}$

expreſ-

expresses the second part of Probability required, which, in the Case of an equality of Skill, would be reduced to $\frac{11}{512}$. Wherefore the Probability required would in this Case be $\frac{3}{256} + \frac{11}{512} = \frac{17}{512}$. Whence it follows, that 'tis a Wager of 495 to 17, or of 29 to 1 very nearly, that in Ten Games, *A* and *B* may not both be in a circumstance of winning, *viz.* *A* the number *q*, and *B* the number *p* of Stakes. But if, by the conditions of the Problem, it were left indifferent whether *A* or *B* should win the Two Stakes or the Three, then the Probability required would be increased, and become as follows, *viz.*

$$\frac{aa+bb}{\overline{a+b}|^2} \times \overline{\frac{a^3 b^3}{\overline{a+b}|^6} + \frac{8a^4 b^4}{\overline{a+b}|^8}}$$

$$\frac{a^3+b^3}{\overline{a+b}|^3} \times \overline{\frac{aabb}{\overline{a+b}|^4} + \frac{7a^3 b^3}{\overline{a+b}|^6}} \, .$$

which, in the Case of an equality of Skill between the Gamesters, would be the double of what it was before.

PROBLEM XLII.

*T*O Find *what Probability there is, that in a given number of Games,* A *may win the number* q *of Stakes; with this farther condition, that* B, *during that whole number of Games, may never have been winner of the number* p *of Stakes.*

SOLUTION.

*F*Rom the Probability that *A* has to win without any limitation the number *q* of Stakes, subtract the Probability there is that both *A* and *B* may be winners, *viz.* *A* of the number *q*, and *B* of the number *p* of Stakes, and there will remain the Probability required.

But, if the conditions of the Problem were extended to this alternative, *viz.* that either *A* should win the number *q* of Stakes, and *B* be excluded the winning of the number *p*; or that *B* should win the number *p* of Stakes, and *A* be excluded the winning of the number *q*, the Probability that either the one or the other of these two Cases may Happen, will easily be deduced from what we have said.

L E M-

LEMMA I.

IN any Series of Terms, whereof the first Differences are equal, the Third Term will be twice the Second, minus once the First; and the Fourth Term likewise will be twice the Third, minus once the Second: Each following Term being always related in the same manner to the two preceding ones. And as this relation is expressed by the two Numbers 2 — 1, I therefore call those Numbers the Index of that Relation.

In any Series of Terms, whose second Differences are equal, the Fourth Term will be three times the Third, minus three times the Second, plus once the First: And each Term in such a Series is always related in the same manner to the three next preceding ones, according to the Index 3 — 3 + 1. Thus, if there be a Series of Squares, such as 4, 16, 36, 64, 100, whose second differences are known to be equal when their Roots have equal Intervals, as they have in this Case, it will be found that the Fourth Term 64 is = $3 \times 36 - 3 \times 16 + 1 \times 4$, and that the Fifth Term 100 is = $3 \times 64 - 3 \times 36 + 1 \times 16$. In like manner, if there were a Series of Triangular numbers, such as 3, 10, 21, 36, 55, whose second Differences are known to be equal, when their sides have equal Intervals, as they have in this Case, it will be found that the Fourth Term is = $3 \times 21 - 3 \times 10 + 1 \times 3$, and that the Fifth Term is = $3 \times 36 - 3 \times 21 + 1 \times 10$; and so on.

So likewise, if there were a Series of Terms whose Third Differences are equal, or whose Fourth Differences are = 0; such as is a Series of Cubes or Pyramidal numbers, or any other Series of numbers generated by the Quantities $ax^3 + bxx + cx + d$, when a, b, c, d being constant Quantities, x is interpreted successively by the Terms of any Arithmetic Progression: Then it will be found that any Term of it is related to the Four next preceding ones, according to the following Index, viz. 4 — 6 + 4 — 1, whose parts are the Coefficients of the Binomial a — b raised to the fourth Power, the first Coefficient being omitted.

And generally, if there be any Series of Terms whose last Differences are = 0. Let the number denoting the rank of that difference be n; then the Index of the Relation of each Term to as many of the preceding ones as there are Units in n, will be expressed by the Coefficients of the Binomial a — b raised to the Power n, omitting the first. But

But if the Relation of any Term of a Series to a constant number of preceding Terms, be expressed by any other Indices than those which are comprised under the foregoing general Law; or even if, those Indices remaining, any of their Signs + or — be changed, that Series of Terms will have none of its differences equal to Nothing.

LEMMA II.

IF in any Series, the Terms A, B, C, D, E, F &c. be continually decreasing, and be so related to one another that each of them may have to the same number of preceding Terms a certain given Relation, always expressible by the same Index; I say, that the sum of all the Terms of that Series ad infinitum may always be obtained.

First, Let the Relation of each Term to the two preceding ones be expressed in this manner, viz. Let C be $=$ m B r — n A rr; and let D likewise be $=$ m C r — n B rr, and so on : Then will the sum of that Infinite Series be equal to $\frac{A + B - mr A}{1 - mr + nrr}$.

Thus, if it be proposed to find the sum of the following Series,

$$\begin{array}{ccccccc} A & B & C & D & E & F & G \end{array}$$

viz. $1 r + 3 rr + 5 r^3 + 7 r^4 + 9 r^5 + 11 r^6 + 13 r^7$ &c. whose Terms are related to one another in this manner, viz. C $=$ 2 r B — 1 rr A, D $=$ 2 r C — 1 rr B &c. Let m and n be made respectively equal to 2 and 1, and these Numerical Quantities being Substituted, in the room of the literal ones, in the general Theorem, the sum of the Terms of the foregoing Series will be found to be equal to $\frac{r + 3 rr - 2 rr}{1 - 2 r + rr}$, or to $\frac{r + rr}{1 - r|^2}$.

Let it be also proposed to find the sum of the following Series

$$\begin{array}{ccccccc} A & B & C & D & E & F & G \end{array}$$

$1 r + 3 rr + 4 r^3 + 7 r^4 + 11 r^5 + 18 r^6 + 29 r^7$ &c. whose Terms are related to one another in this manner, viz. C $=$ 1 B r + 1 A rr, D $=$ 1 C r + 1 B rr &c. Let m and n be respectively made equal to 1 and — 1, and then that Series will be found equal to $\frac{r + 3 rr - rr}{1 - r - rr}$, or to $\frac{r + 2 rr}{1 - r - rr}$.

DEMON.

DEMONSTRATION.

Let the following Scheme be written down, *viz.*

$$A = A$$
$$B = B$$
$$C = m\,Br - n\,Arr$$
$$D = m\,Cr - n\,Brr$$
$$E = m\,Dr - n\,Crr$$
$$F = m\,Er - n\,Drr$$
&c.

This being done, if the sum of the Terms A, B, C, D, E, F &c. *ad infinitum*, compofing the firft Column, be fuppofed equal to x, then the fum of the Terms of the other two Columns will be found thus: By *Hypothefis*, $A + B + C + D + E$ &c. $= x$, or $B + C + D + E$ &c. $= x - A$; and Multiplying both fides of this Equation by mr, it will follow that $m\,Br + m\,Cr + m\,Dr + m\,Er$ &c. is $= mrx - mrA$. Again, adding $A + B$ on both fides, we fhall have the fum of the Terms of the fecond Column, *viz.* $A + B + m\,Br + m\,Cr + m\,Dr$ &c. equal to $A + B + mrx - mrA$. The fum of the Terms of the third Column will be found by bare infpection to be $- nrrx$. But the fum of the Terms contain'd in the firft Column, is equal to the other two fums contained in the other two Columns. Wherefore the following Equation will be had, *viz.* $x = A + B + mrx - mrA - nrrx$; from whence it follows that the value of x, or the fum of all the Terms $A + B + C + D + E$ &c. will be equal to

$$\frac{A + B - mrA}{1 - mr + nrr}$$

Secondly, Let the Relation of each Term to the three next preceding ones be expreffed as follows, *viz.* let D be $= m\,Cr - n\,Brr + p\,Ar^3$, and let E likewife be $= m\,Dr - n\,Crr + p\,Br^3$, and fo on: Then will the fum of all the Terms $A + B + C + D + E$ &c. *ad infinitum*, be equal to

$$\frac{A + B + C - mrA + nrrA - mrB}{1 - mr + nrr - pr^3}$$

L l

To apply this Theorem, let it be proposed to find the sum of the following Series,

$$A \quad\quad B \quad\quad C \quad\quad D \quad\quad E \quad\quad F \quad\quad G$$
$$4r + 16rr + 36r^3 + 64r^4 + 100r^5 + 144r^6 + 196r^7 \text{ \&c.}$$

whose Numerical Quantities are related to one another according to the Index $3 - 3 + 1$, corresponding to $m - n + p$: Let therefore $3, 3, 1$ be substituted in the room of m, n, p; let also $4r, -16rr, 36r^3$ be substituted in the room of A, B, C: Then the sum of the Terms of the foregoing Series will be found equal to $\dfrac{4r + 16rr + 36r^3 - 12rr - 48r^3 + 12r^3}{1 - 3r + 3rr - r^3}$, or

$$\frac{4r + 4rr}{1 - r|^3}.$$

And in like manner the sum of the Terms of the following Series, *viz.*
$$r + 2rr + 5r^3 + 20r^4 + 72r^5 + 261r^6 + 947r^7 \text{ \&c. }$$ whose Numerical Quantities are related to one another according to the Index $3 + 2 + 1$, will by a proper substitution, be found to be equal to $\dfrac{r - rr - 3r^3}{1 - 3r - 2rr - r^3}$.

Thirdly, Let the Relation of each Term of a Series to four of the preceding Terms be expressed by means of the Index: $m - n + p - q$, and the Sum of that Series will be

$$\frac{A + B + C + D - mrA + nrrA - pr^3 A}{- mrB + nrrB}$$
$$\frac{- mrC}{1 - mr + nrr - pr^3 + qr^4}$$

Fourthly, The Law of the continuation of these Theorems being manifest, they may be all easily comprehended under one general Rule.

Fifthly, If the corresponding Terms of any two or more Series, generated after the manner which we have above described, be multiplyed by one another, the new Series resulting from that multiplication, will also be exactly summable: Thus, taking the two following Series, *viz.*

$$r + 2rr + 3r^3 + 5r^4 + 8r^5 + 13r^6 \text{ \&c.}$$
$$r + 3rr + 4r^3 + 7r^4 + 11r^5 + 18r^6 \text{ \&c. in both of}$$
$$\text{which.}$$

which each Numerical Quantity is the fum of the two preceding ones; the Series refulting from the multiplication of the corresponding Terms will be

$$rr + 6rr^4 + 12r^6 + 35r^8 + 88r^{10} + 234r^{12} \text{ &c.}$$

in which each Numerical Quantity being related to the three preceding ones, according to the Index $2 + 2 - 1$, the fum of that Series will be found to be $= \frac{rr + 4r^4 - 2r^6}{1 - 2rr - 2r^4 + r^6}$

as will appear, if in the room of $m - n + p$ there be fubftituted $2 + 2 - 1$, and rr be written inftead of r.

When the Numerical Quantities belonging to the Terms of any Series are reftrained to have their laft differences equal to Nothing, then may the fums of thofe Series be alfo found by the following elegant Theorem, which has been communicated to me by *Mr. de Monmort.*

Let Ar be the firft Term of the Series, and let the firft, fecond and third differences, &c. of the Numerical Quantities belonging to the Terms of the Series, be refpectively equal to d' d'', d''', &c. Then will the fum of the Series be equal to

$$\frac{Ar}{1-r} + \frac{rd'}{1-r|^2} + \frac{r^2 d''}{1-r|^3} + \frac{r^3 d'''}{1-r|^4} + \frac{r^4 d''''}{1-r|^5} \text{ &c.}$$

Thus, if it were propofed to find by this Theorem the fum of the following Series, *viz.*

$$Ar + 16rr + 36r^3 + 64r^4 + 100r^5 \text{ &c.}$$

It is plain that in this cafe A is $= 4$, $d' = 12$, $d'' = 8$, $d''' = 0$; and therefore that the fum of this Series is equal to

$$\frac{4r}{1-r} + \frac{12rr}{1-r|^2} + \frac{8r^3}{1-r|^3}, \text{ which is reduced to } \frac{4r+4rr}{1-r|^3}.$$

REMARK.

OUr Method of fumming up all the Terms which in thefe Series are related to one another according to conftant Indices, may be extended to the finding of the fum of any determinate number of thofe Terms. Thus, if A, B, C, D be the firft Terms of a Series, and U, X, Y, Z be the laft, then will the fum of the Series be

A —

$$
\begin{array}{l}
A - mr\,A + nrr\,A - pr^3\,A + qr^4\,U \\
+\, B - mr\,B + nrr\,B - pr^3\,X + qr^4\,X \\
+\, C - mr\,C + nrr\,T - pr^3\,T + qr^4\,T \\
+\, D - mr\,Z + nrr\,Z - pr^3\,Z + qr^4\,Z \\
\hline
1 - mr\; +\; nrr\; -\; pr^3\; +\; qr^4
\end{array}
$$

And if a general Theorem were defired, it might eafily be formed from the infpection of the foregoing.

These Theorems are very ufeful for fumming up readily thofe Series which exprefs the Probability of the Plays being Ended in a given number of Games. For example, fuppofe it be required to find what Probability there is, that in Four and twenty Games, either A fhall win Four Stakes of B, or B Four Stakes of A. The Series expreffing that Probability is, from our XXXIV*th* Problem

$$
-\frac{a^4+b^4}{a+b^4}\times \overline{\; 1 + \frac{4ab}{a+b^2} + \frac{14aabb}{a+b^4} + \frac{48\,a^3b^3}{a+b^6} + \frac{164\,a^4b^4}{a+b^8}\;} \quad \&c.
$$

or, fuppofing an equality of Skill between the two Game-fters, $\frac{1}{8}\times\; \overline{1 + \frac{4}{4} + \frac{14}{16} + \frac{48}{64} + \frac{164}{256}}\; \&c.$ which ought to be continued to eleven Terms independently from the common Multiplicator. Let this Series, whofe Terms are related according to the Index $4-2$, be compared with the Theorem, making $A = 1$, $B = \frac{4}{4} = 1$, $m = 4$, $n = 2$. and neglecting the Terms C, D, U, X, the fum of the aforefaid Series will be found $= 8 + T - 7\,Z$; which being multiplied by the common Multiplicator $\frac{1}{8}$ prefixt to it, the Probability required will be expreffed by $1 + \frac{1}{8}T - \frac{7}{8}Z$. Wherefore nothing remains to be done but to find the two laft Terms T and Z: But thofe two Terms, by our XXXIV*th*. Problem, will be found to be $\frac{76096}{2^{18}}$ and $\frac{159808}{2^{20}}$, or 0.2902, and 0.2477 nearly; which numbers being fubftituted refpectively in the room of T and Z, the Probability required will be found to be equal to 0.8193 nearly. Let now this laft number be fubtracted from Unity, and the remainder being 0.1807, it follows, that 'tis a Wager of 82 to 18, or of 41 to 9 nearly, that in Twenty four Games or fooner, either A fhall win four Stakes of B, or B four Stakes of A.

If

If the number of Stakes were Five, the fum of the Terms of the Series belonging to that Cafe would alfo be expreft by means of the two laft Terms, fuppofing any given number of Games, or any proportion of Skill.   If the number of Stakes were Six or Seven, the fum of the Series belonging to thofe Cafes would be expreft by means of the three laft Terms; If Eight or Nine, by means of the Four laft Terms, and fo on.

LEMMA III.

IF there be a Series of Numbers, as A, B, C, D, E &c. *whofe Relation is expreft by any conftant Index, and there be another Series of Numbers, as* P, Q, R, S, T &c. *whofe laft Differences are equal to Nothing; and each Term of the firft Series be Multiplied by each correfponding Term of the fecond, I fay that the Products* AP, BQ, CR, DS, ET &c. *conftitute a Series of Terms, whofe Relation may be expreft by a conftant Index. Thus if we take the Series* 1, 2, 8, 28, 100 &c. *whofe Terms are related by the Index* 3+2, *and each Term of that Series be refpectively Multiplied by the correfponding Terms of an Arithmetic Progreffion, fuch as* 1, 3, 5, 7, 9 &c. *whofe laft Differences are equal to Nothing: Then it will be found that the Products* 1, 6, 40, 196, 900 &c. *conftitute a Series of Numbers, each Term of which is Related to the preceding ones according to the Index* 6 — 5 — 12 — 4. *Now the Rule for finding the Index of this Relation is as follows.*

Take the Index which expreffes the Relation of the Terms in the firft Series, and Multiply each Term of it by the correfponding Terms of the Literal Progreffion r, rr, r³ &c. *which being done, fubtract the fum of thefe Products from Unity; then let the remainder be raifed to its Square, if the fecond Series be compofed of Terms in Arithmetic Progreffion; or to its Cube, if it be compofed of Terms whofe third Differences are equal to Nothing; or to its fourth Power, if it be compofed of Terms whofe fourth Differences are equal to Nothing; and fo on. Let that Power be fubtracted from Unity, and the remainder, having cancelled the Letter* r, *will be the Index required. Thus in the foregoing Example, having taken the Index* 3+2, *which belongs to the firft Series, and Multiplied its Terms by* r *and* rr *refpectively, let the Product* 3r + 2rr be *subtracted from Unity, and the Square of the remainder being* $1 - 6r + 5rr + 12r^3 + 4r^4$, *let that Square be also subtracted from Unity, then the remainder, having cancelled the Letter* r, *will be* $6 - 5 - 12 - 4$, *which is the Index required.*

But *in case neither of the two first Series have any of their last Differences equal to Nothing, yet if in both of them the Relation of their Terms be expressed by constant Indices, the third Series, resulting from the Multiplication of the corresponding Terms of the two first Series, will also have its Terms related to one another according to a constant Index. Thus, taking the Series* 1, 3, 5, 11, 21, 43 &c. *the Relation of whose Terms is expressed by means of the Index* $1 + 2$, *and Multiplying its Terms by the corresponding Terms of the Series* 1, 2, 5, 13, 34, 89 &c. *the Relation of whose Terms is expressed by the Index* $3 - 1$, *the Products will compose the Series* 1, 6, 25, 143, 714, 3827, *whose Terms are Related to one another according to the Index* $3 + 13 - 6 - 4$.

Generally, *If the Index expressing the Relation of the Terms in the first Series be* $m + n$, *and the Index expressing the Relation of the Terms in the second Series be* $p + q$; *then will the Index of the Relation, in the Series resulting from the Multiplication of the corresponding Terms of the Two first Series, be expressed by the following Quantities,*

$$viz. \quad mp + \begin{array}{c} + mmq \\ npp \\ + 2nq \end{array} + mnpq - nnqq.$$

But if it *so Happen that* p *be equal to* m, *and* q *to* n; *then the foregoing Theorem may be contracted, and the Index of the Relation may be expressed as follows, viz.* $\dfrac{mm + mmn}{+ n + nn} - n^3$

so that the Relation of each Term to the preceding ones need not be extended, in this Case, to any more than three Terms.

And *in like manner other Theorems may be found, which may be extended farther, and at last be comprized under one general Rule.*

P R O.

PROBLEM XLIII.

Suppoſing A *and* B, *whoſe proportion of Skill is as* a *to* b, *to Play together, till* A *either wins the number* q *of Stakes, or loſes the number* p *of them; and that* B *Sets at every Game the ſum* G *to the ſum* L: *It is required to find the Advantage, or Diſadvantage of* A.

SOLUTION.

First, Let the number of Stakes to be won or loſt on either ſide be equal, and let that number be *p*; let there be alſo an equality of Skill between the Gameſters: Then I ſay, the gain of *A* will be $pp \times \frac{G-L}{2}$, that is, the Square of the number of Stakes which either Gameſter is to win or loſe, Multiplied by one half of the Difference of the value of the Stakes. Thus, if *A* and *B* play till ſuch time as Ten Stakes are won or loſt, and *B* Setts a Guinea to Twenty Shillings; then the gain of *A* will be a hundred times half the Difference between a Guinea and Twenty Shillings, *viz.* 3 *l* — 19 ˢʰⁱˡ.

Secondly, Let the number of Stakes be unequal, ſo that *A* be obliged either to win the number *q* of Stakes, or to loſe the number *p*; let there be alſo an equality of Chance between *A* and *B*: Then I ſay, that the gain of *A* will be $pq \times \frac{G-L}{2}$; that is, the Product of the two numbers of Stakes, and one half of the Difference of the value of the Stakes Multiplied together. Thus, if *A* and *B* play together till ſuch time as either *A* wins Eight Stakes, or loſes Twelve; then the Gain of *A* will be the Product of the Three numbers 8, 12, 9, which makes 864 pence, or 3 *l* — 12 ˢʰⁱˡ.

Thirdly, Let the number of Stakes be equal, but let the number of Chances to win a Game, or the Skill of the Gameſters be unequal, in the proportion of *a* to *b*. Then I ſay,

that the gain of *A* will be $\frac{\overline{pa^p - pb^p}}{a^p + b^p} \times \frac{aG - bL}{a - b}$

Fourthly,

Fourthly, Let the number of Stakes be unequal, and let also the number of Chances be unequal: Then I say that the gain of A will be

$$\frac{\overline{q\,a^q \times a^p - b^p} - p b^p \times \overline{a^q - b^q}}{a^{p+q} - b^{p+q}} \times \frac{\overline{a\,G - b L}}{a - b}$$

DEMONSTRATION.

IN order to form a general Demonstration of these Rules, let us resolve some particular Cases of this Problem, and examine the procefs of their Solution: Let it therefore be propofed to find the gain of A in the Cafe of Four Stakes to be won or loft on either fide, and of an equality of Chance between A and B to win a Game. There being an equality of Chance for A, every Game he plays, to win G or to lofe L, it follows, that the gain of every Game he plays is to be reputed to be $\frac{G-L}{2}$. But it being uncertain whether any more Games than Four will be play'd, it follows, that the gain of the Tenth Game, for inftance, to be eftimated before the play begins, cannot be reputed to be $\frac{G-L}{2}$; for it would only be fuch provided the Play were not Ended before that Tenth Game: Wherefore the gain of the Tenth Game is the Quantity $\frac{G-L}{2}$ Multiplied by the Probability of the Plays not being Ended in Nine Games, or before, for the fame reafon, the gain of the Ninth Game is the Quantity $\frac{G-L}{2}$ Multiplied by the Probability that the Play will not be Ended in Eight Games: And likewife the gain of the Eighth Game is the Quantity $\frac{G-L}{2}$ Multiplied by the Probability that the Play will not be ended in Seven Games, and fo on. From whence it may be concluded, that the gain of A, to be eftimated before the Play begins, is the Quantity $\frac{G-L}{2}$ Multiplied by the fum of the Probabilities that the Play will not be Ended in 0, 1, 2, 3, 4, 5, 6, &c. Games *ad infinitum*.

Let thofe Probabilities be refpectively called A', B', C', D', E', F', G', &c. Then, becaufe the Probability of the Plays not being Ended in Five Games is equal to the Probability of its not Ending in Four, and that the Probability of its not Ending in Seven, is equal to the Probability of its not

Ending

Ending in Six, it will follow, that the sum of the Probabilities belonging to all the Even Games is equal to the sum of the Probabilities belonging to all the Odd ones : We are therefore only to find the sum of all the Even Terms, $A' + C' + E' + G'$ &c. and to double it afterwards.

Now it will appear, from our XXXIII*d* Problem, that these Terms constitute the following Series, *viz.*

$$\frac{1}{1} + \frac{4}{4} + \frac{14}{16} + \frac{48}{64} + \frac{164}{256} + \frac{560}{1024} + \frac{1912}{4096} \&c.$$

In which Series, each Numerator being Related to the two preceding ones according to the Index $4 - 2$, and each Denominator being a Power of 4, it follows, that this Series may be compared with the first Theorem of our second *Lemma*; by making the first and second Terms A and B, used in that Theorem, to be respectively equal to 1 and $\frac{4}{4}$, making also the Quantities m, n, r respectively equal to 4, 2, $\frac{1}{4}$. Which being done, it will be found that the sum of all those Terms *ad infinitum* will be equal to 8.

We may therefore conclude that the sum of all the Terms $A' + B' + C' + D' + E'$ &c. is equal to 16, and that the gain of A is equal to $16 \times \frac{G - L}{2}$.

But if the number of Chances which A and B have to win a Game, be in a proportion of inequality, then the sum of the Series $A' + C' + E' + G' + I'$ &c. will be found thus : Let $\frac{ab}{a+b|^2}$ be called r, and the Terms of that Series will be Related to one another as follows, *viz.* $E' = 4 C'r - 2 A'rr$, $G' = 4 E'r - 2 C'rr$, and so on. Let therefore 4, 2, 1, 1, be respectively substituted, in the first Theorem of our second *Lemma*, in the room of m, n, A', B'; and the sum of this Series will be found to be $\frac{2 - 4r}{1 - 4r + 2rr}$; in which expression, restoring the value of r, *viz.* $\frac{ab}{a+b|^2}$, the sum of the Series will become $\frac{2aa + 2bb \times \overline{a+b}|^2}{a^4 + b^4}$, the double of which is the sum of all the Terms $A' + B' + C' + D' + E'$ &c. But because, in every Game, the Gamester A has the number a of Chances to win G, and the number b of Chances to lose L; it follows, that his gain in every Game is equal

N n to

to $\frac{aG-bL}{a+b}$. From whence it may be concluded, that the Advantage of A, to be estimated before the Play begins,

will be $\overline{\frac{4aa+4bb \times \overline{a+b}^2}{a^4+b^4}} \times \frac{aG-bL}{a+b}$.

Before we proceed farther, we must observe, that the Series $A' + B' + C' + D' + E' + F'$ &c. which we have assumed to represent in general the Probabilities of the Plays not being Ended in 0, 1, 2, 3, 4, 5 &c. Games, whether the Stakes be equal or unequal, being divided into two parts, *viz.* $A' + C' + E' + G'$ &c. and $B' + D' + F' + H'$ &c. answering to 0, 2, 4, 6 &c. and 1, 3, 5, 7, &c. each Term of these two new Series will be related to the preceding ones, according to the same Law of Relation, as are the Terms of those Series which express the Probabilities of the Plays being Ended in a certain number of Games, under the like circumstances of Stakes to be won or Lost. The Law of which Relation is to be deduced from our XXXIV*th*, and XXXIX*th* Problems.

If the number of Stakes to be won or lost on either side be equal to Six, and the proportion of Chances to win a single Game be as a to b; then the Relation of each Term to the preceding ones, in the Series $A' + C' + E' + G'$ &c. will be expressed by the Index $6 - 9 + 2$. Wherefore to find the sum of these Terms, let the Quantities 6, 9, 2, 1, 1, 1, be respectively substituted, in the third Theorem of our second *Lemma*, in the room of m, n, p, A', B', C', and the sum of those Terms will be found to be

$$\frac{1+1+1-6r+9rr}{-6r}\Big/(1-6r+9rr-2r^3) \text{ or } \frac{3-12r+9rr}{1-6r+9rr-2r^3}.$$ In which expression substituting $\frac{ab}{\overline{a+b}^2}$ in the room of r, the same will

become $\overline{\frac{3a^4+3aabb+3b^4 \times \overline{a+b}^2}{a^6+b^6}}$. From whence we may

conclude, that the gain of A will be

$$\overline{\frac{6a^4+6aabb+6b^4 \times \overline{a+b}^2}{a^6+b^6}} \times \frac{aG-bL}{a+b}$$

Again,

Again, if the number of Stakes to be won or loft on either fide be Eight, it will be found, that the gain of *A* will

be $\overline{\frac{8a^6 + 8a^4bb + 8aab^4 + 8b^6 \times a + \beta^2}{a^8 + b^8}} \times \overline{\frac{aG - bL}{a + b}}$.

But the Numerators of the foregoing Fractions being in Geometric Progreffion, if thofe Progreffions be fummed up, the gain of *A*, in the Cafe of Four Stakes to be won or loft, may be expreffed as follows,

viz. by the Fraction $\overline{\frac{4a^4 - 4b^4 \times a + \beta^2}{a^4 + b^4 \times aa - bb}} \times \overline{\frac{aG - bL}{a + b}}$; or

dividing both Numerator and Denominator by $\overline{a + \beta}^2$, the fame may be expreft by the Fraction $\overline{\frac{4a^4 - 4b^4}{a^4 + b^4}} \times \overline{\frac{aG - bL}{a - b}}$. His gain likewife, in the Cafe of Six Stakes to be won or loft, will be expreft by the Fraction $\overline{\frac{6a^6 - 6b^6}{a^6 + b^6}} \times \overline{\frac{aG - bL}{a - b}}$; and in the Cafe of Eight Stakes to be won or loft, it will be expreft by the Fraction $\overline{\frac{8a^8 - 8b^8}{a^8 + b^8}} \times \overline{\frac{aG - bL}{a - b}}$: So that we may conclude, that in any Cafe of an Even and equal number of Stakes denominated by *p*, the gain of *A* will be expreft by the Fraction $\overline{\frac{pa^p - pb^p}{a^p + b^p}} \times \overline{\frac{aG - bL}{a - b}}$.

But if the number of Stakes be Odd and equal, as it is in the Cafe of Five Stakes to be won or loft, then the two Series $A' + C' + E' + G' + I'$ &c. and $B' + D' + H'$ &c. will be unequal, and the excefs of the firft above the fecond will be Unity. Wherefore to find the gain of *A*, in the Cafe of Five Stakes, having fet afide the firft Term of the firft Series, let all other the Terms be added together, by comparing them with thofe that are employed in the firft Theorem of our fecond *Lemma*; which will be done thus. Since $C' = 1$, $E' = 1$, and $G' = 5 E'r - 5 C'rr$, let the numbers 1, 1, 5, 5, be refpectively fubftituted in the aforefaid Theorem, in the room of the Letters A', B', *m*, *n*;

and

and the sum of that Series will be found to be $\frac{2 - 5r}{1 - 5r + 5rr}$:
To the double of which adding Unity, which we had set aside, it will appear that the sum of the two Series together will be $\frac{5 - 5r + 5rr}{1 - 5r + 5rr}$; or writing $\frac{ab}{a+b^2}$ in the room of r,
$\frac{5a^4 + 5a^3b + 5aabb + 5ab^3 + 5b^4}{a^4 - a^3b + aabb - ab^3 + b^4}$. Now by reason that the Terms of both Numerator and Denominator of this last Fraction compose a Geometrick Progression, the Numerator will be reduced to $\frac{5a^5 - 5b^5}{a - b}$, and the Denominator will be reduced to $\frac{a^5 + b^5}{a + b}$. From whence it follows, that the sum of these two Series will be $\frac{\overline{5a^5 - 5b^5} \times \overline{a+b}}{a^5 + b^5 \times a - b}$, and that the gain of A will be $\frac{\overline{5a^5 - 5b^5}}{a^5 + b^5} \times \frac{\overline{aG - bL}}{a - b}$. If the gain of A be likewise inquired into, in the Case of Seven Stakes to be won or lost, then it will be found to be $\frac{\overline{7a^7 - 7b^7}}{a^7 + b^7} \times \frac{\overline{aG - bL}}{a - b}$. And the same form of expression being constantly observed in all cases wherein the number of Stakes is Odd and equal, we may conclude that if that number be denominated by p, then the gain of A will be $\frac{\overline{p\,a^p - p\,b^p}}{a^p + b^p} \times \frac{\overline{aG - bL}}{a - b}$. Now this expression of the gain of A having been found to be the same in the Case of an Even number of Stakes, as it is now found in the Case of an Odd one; we may conclude, that it is general, and belongs to any equal number of Stakes whether Even or Odd.

If the number of Stakes be unequal, the Investigation of the gain of A will be made in the same manner as it was in the Case of an equality of Stakes. Thus, let us suppose that the Play be to continue till such time as either A wins Two Stakes, or B Three. In order therefore to find the gain of A, let the Series $A' + B' + C' + D' + E' + F'$ &c. be be divided into two parts, *viz.* $A' + C' + E'$ &c. and $B' + D' + F'$ &c. then it will appear, from our XXXIII*d*, and XXXIX*th* Problems, that $A' = 1$, $C' = \frac{2ab + bb}{a + b|^2}$, $E' = 3\,C'r - 1\,A'\,rr$. Having now obtained the firſt Terms of the Series, and the Relation of each Term of it to the preceding ones; it will be eaſie to find the ſum of all its Terms, by the help of the firſt Theorem of our ſecond *Lemma*, making the Quantities, A, B, m, n therein employed to be reſpectively equal to 1, $\frac{2ab + bb}{a + b|^2}$, 3, 1. This done, the ſum of that Series will be found to be equal to

$$\frac{\overline{aa + ab + 2bb} \times \overline{a + b|^2}}{a^4 + a^3b + aabb + ab^3 + b^4}.$$

In like manner it will appear, that in the ſecond Series B' is $= 1$, $D' = \frac{2aab + 3abb}{a + b|^2}$; $F' = 3\,D'r - B'\,rr$; from whence the ſum of all its Terms will be found to be $\frac{\overline{aa + 2ab + 2bb} \times \overline{a + b|^2}}{a^4 + a^3b + aabb + ab^3 + b^4}$: And both ſums of thoſe Series being added together, the aggregate of them will be $\frac{\overline{2a^3 + aab + 6abb + 3b^3} \times \overline{a + b}}{a^4 + a^3b + aab + ab^3 + b^4}$: But the Terms of this Denominator compoſing a Geometric Progreſſion, whoſe ſum is $\frac{a^5 - b^5}{a - b}$, the foregoing Fraction may be reduced to $\frac{\overline{2a^4 + 2a^3b - 2aabb - 3ab^3 - 3b^4} \times \overline{a + b}}{a^5 - b^5}$; which Fraction is ſtill capable of a farther reduction; for the three firſt Terms of its Numerator compoſe a Geometric Progreſſion, and the two laſt Terms may be conſidered as being in Geometric Progreſſion, and conſequently the Fraction may at laſt be reduced to $\frac{2aa.\,\times\,\overline{a^3 - b^3} - 3b^3 \times \overline{aa - bb} \times \overline{a + b}}{a^3 - b^3 \times \overline{a - b}}$, from which expreſſion, the gain of A will be found to be

$$\frac{2aa \times \overline{a^3 - b^3} - 3b^3 \times \overline{aa - bb}}{a^5 - b^5} \times \frac{aG - bL}{a - b}.$$

By

By the fame method of Procefs, it will be eafy to deter-mine the gain of *A* under any other circumftance of Stakes to be won or loft: And if it be remembred always to fum up thofe Terms which are in Geometric Progreffion, all the various expreffions of the gain of *A*, calculated for differing numbers of Stakes, will appear to be uniform: From whence it may be collected by bare infpection, that the gain of *A* is what we have afferted it to be, *viz.*

$$\overline{\frac{q a q \times a^p - b^p - p\, b^p \times a^q - b^q}{a^{p+q} - b^{p+q}}} \times \overline{\frac{a\,G - b\,L}{a - b}}.$$

It is to be obferved, *Firft,* that if *p* and *q* be equal, the fore-going expreffion may be reduced to $\overline{\frac{p a^p - p b^p}{a^p + b^p}} \times \overline{\frac{a\,G - b\,L}{a - b}}$, as will appear if both Numerator and Denominator be divi-ded by $a^p - b^p$, having firft fubftituted *p* in the room of *q*. *Secondly,* that if *a* and *b* be equal, the fame expreffion may be reduced to $p\, q \times \overline{\frac{G - L}{2}}$, which will appear if both Nume-rator and Denominator be divided by $\overline{a - b}^2$.

After I had Solved the foregoing Problem, I wrote word of it to Mr. *Nicolas Bernoully,* the prefent Profeffour of Ma-thematics at *Padoua,* without acquainting him with my So-lution: I only let him know in general that it was done by the Method of Infinite Series; whereupon he fent me two different Solutions of that Problem: And as one of them has fome Affinity with the Method of Series ufed all along in this Book, I fhall tranfcribe it here in the Words of his Let-ter, " My Uncle has obferved that this Problem may alfo " be Solved after the fame manner as you have Solved the " Ninth Problem * of your Tract *de Menfura Sortis,* it be-" ing vifible that the Expectations of the Gamefters will re-" ceive no alteration whether it be fuppofed that the Pieces " which *A* and *B* Set every time to each other, are refpective-" ly *L* and *G,* or whether it be fuppofed that thofe Pieces " conftitute the following Progreffion, *viz.*

" $L, G, G + \frac{b}{a} \times \overline{G - L}, G + \frac{b}{a} \times \frac{b b}{a a} \times \overline{G - L}, G + \frac{b}{a} \times \frac{b b}{a a} \times \frac{b^b}{a^b}$ $\times \overline{G - L}$ &c. the number of whole Terms is *p* + *q.* whereof

" the firſt, whoſe number is p, denote the Pieces of A; and
" the laſt, whoſe number is q, denote the Pieces of B: For in
" either Caſe the gain of A will be $\frac{aG - bL}{a + b}$. Now it being
" poſſible to find the ſum of any number of Terms of this Pro-
" greſſion, it follows that the different values of all the Pieces
" of each Gameſter may be obtained: Let therefore thoſe
" values be denoted reſpectively by S and T; let alſo the
" Probabilities of winning the number of Stakes agreed up-
" on be called A and B reſpectively, which Probabilities are

" $\frac{a^{p+q} - a^q b^p}{a^{p+q} - b^{p+q}}$ and $\frac{a^q b^p - b^{p+q}}{a^{p+q} - b^{p+q}}$, ſuch as we had ſeverally
" derived them, your ſelf in your aforeſaid Problem, and I
" in Mr. *Montmort*'s Book. This being ſuppoſed, the gain of A
" will be found to be $AT - BS$, or $\frac{aG - bL}{a - b} \times \overline{Aq - Bp}$.

N. B. Tho' I may, accidentally, have given a uſeful Hint
for that elegant Method of ſolving the foregoing Problem, yet
I think it reaſonable to aſcribe it entirely to its proper Author;
the Hint having been improv'd much beyond what I could
have expected.

REMARK.

IT is to be obſerved, that the gain of A is not to be regu-
lated by the equal Probability there is that the Play may,
or may not be Ended in a certain number of Games. For
inſtance, If two Gameſters having the ſame number of Chances to win a Game, deſign only to play untill ſuch time only
as two Stakes are won or loſt; it is as Probable that the Play
may be Ended in two Games as not, yet it cannot be concluded from thence, that the gain of A is to be eſtimated
by the Product of the Number 2 by one half of the Difference of the Stakes: For it has been Demonſtrated that this
gain will be Four times that half difference. In like manner,
if the Play were to continue, till either A ſhould win Two
Stakes, or B Three; it will be found, that it is as Probable
that the Play may End in Four Games as not; and yet the
gain of A is not to be eſtimated by the Product of the Number 4 by one half of the Difference of the Stakes; it having

ving been Demonſtrated that it is Six times that half Difference. To make this the more ſenſible, let us ſuppoſe that
A and *B* are to Play till ſuch time as *A* either wins one
Stake, or loſes Ten: It is plain, that in this Caſe it is as Probable that the Play may be Ended in One Game as not, and
yet the gain of *A* will be found to be Ten times the Difference of the Stakes. From hence it is plain, that this gain
is not to be eſtimated, by the equal Probability of the Plays
Ending or not Ending in a certain number of Games, but
by the Rules which have been preſcribed in this Problem.

PROBLEM XLIV.

IF A *and* B, *whoſe proportion of Skill is as* a *to* b, *reſolving
to Play together till ſuch time as Four Stakes are won or loſt
on either ſide, agree between themſelves, that the firſt Game that
is play'd, they ſhall Set to each other the reſpective ſums* L *and* G;
that the ſecond Game they ſhall Set the ſums 2 L *and* 2 G; *the
third Game the ſums* 3 L *and* 3 G, *and ſo on; the Stakes increa-
ſing continually in an Arithmetic Progreſſion : It is Demanded how
the gain of* A *is to be eſtimated in this Caſe, before the Play
begins.*

SOLUTION.

LET there be ſuppoſed a Time wherein the Number p
of Games has been play'd; then *A* having the Num-
ber a of Chances to win the ſum $\overline{p+1} \times G$ in the next Game,
and *B* having the Number b of Chances to win the ſum
$\overline{p+1} \times L$; it is plain, that the gain of *A* in that circumſtance
of Time will be $\overline{p+1} \times \frac{aG-bL}{a+b}$. But this gain being to be
eſtimated before the Play begins, it follows, that it ought to
be eſtimated by the Quantity $\overline{p+1} \times \frac{aG-bL}{a+b}$ multiplied by
the reſpective Probability there is that the Play will not
then be Ended ; and therefore the whole gain of *A* is the
ſum of the Probabilities of the Plays not Ending in 0, 1, 2,
3, 4, 5, 6 &c. Games *ad infinitum*, multiplied by the reſpect-
ive values of the Quantity $\overline{p+1} \times \frac{aG-bL}{a+b}$, p being Interpreted

preted fucceffively by the Terms of the Arithmetic Progreffion, 0, 1, 2, 3, 4, 5, 6 &c. Now let thefe Probabilities of the Plays not Ending be refpectively called A', B', C' D', E', F', G' &c. Let alfo the Quantity $\frac{aG - bL}{a+b}$ be called S; and thence it will follow, that the gain of A will be $A'S + 2B'S + 3C'S + 4D'S + 5E'S + 6F'S$ &c. But in the Cafe of this Problem B' is equal to A', and D' is equal to C', and fo on. Wherefore the gain of A may be expreffed by the Series $S \times \overline{3A' + 7C' + 11E' + 15G' + 19F}$ &c. But it appears, by our XXXIIId Problem, that the Terms A', C' E', G' are refpectively equal to the following Quantities, *viz.* 1, 1, $\dfrac{4a^3b + 6aabb + 4ab^3}{a+b|^4}$,

$\dfrac{14a^4bb + 20a^3b^3 + 14aab^4}{a+b|^6}$: Whence it follows, that

the Terms $3A' + 7C' + 11E' + 15G'$ may be obtained: It appears alfo, from what we have obferved in the preceding Problem, that the Relation of the Terms A', C', E' &c. may be expreffed by the Index $4 - 2$; and by the Third *Lemma* prefixt to that Problem, that the Relation of the Terms $3A'$, $7C'$, $11E'$ &c. may be expreffed by the Index $8 - 20 + 16 - 4$: And therefore fubftituting the Quantities $3A'$, $7C'$, $11E'$, $15G'$ in the room of the Quantities A, B, C, D, which we make ufe of in the Third Theorem of our fecond *Lemma*; fubftituting likewife the Quantities, 8, 20, 16, 4 in the room of m, n, p, q; and laftly fubftituting $\frac{ab}{a+b|^2}$ in the room of r; the gain of A will be expreft by the following Quantities, *viz.*

$$S \times \frac{\overline{10a^6 + 24a^5b + 42a^4bb + 64a^3b^3 + 42aab^4 + 24ab^5 + 10b^6} \times \overline{a+b}|^4}{\overline{a^4 + b^4}|^2}$$

which, in the Cafe of an equal number of Chances to win a Stake, would be reduced to $216 S$; and therefore if the Quantities G and L ftand refpectively for a Guinea and Twenty Shillings, which will make the value of S to be Nine pence, it follows, that the gain of A will in this Cafe be $8 l - 2^{\text{shil.}}$

Corollary I.

Corollary I. If the Stakes were to Increafe according to the proportion of the Terms of any of thofe Series which we have defcribed in our *Lemma's,* and that there were any given inequality in the number of Stakes to be won or loft, the gain of *A* might ftill be found.

Corollary II. There are fome Cafes wherein the gain of *A* would be Infinite: Thus, if *A* and *B* were to Play till fuch times as Four Stakes were won or loft, and it were agreed between them to double their Stakes at every Game, the gain of *A* would in this Cafe be Infinite: Which confequence may eafily be deduced from what has been faid before.

PROBLEM XLV.

IF A *and* B *refolve to Play till fuch time as* A *either wins a certain given number of Stakes, or that* B *wins the fame, or fome other given number of them:* 'Tis required to find in how many Games it will be as Probable that the Play may be Ended as not?

SOLUTION.

LET it be fuppofed that *A* and *B* are to play till fuch time as either of them wins Three Stakes, and that there is an Equality of Skill between them. This being fuppofed, it will appear, from our XXXIV*th* Problem, that the Probability of the Plays continuing for an Indeterminate number of Games may be expreft by the following Series, *viz.*

$$\frac{\overline{a^3 + b^3}}{\overline{a + b}^3} \times \overline{1 + \frac{3ab}{\overline{a + b}^2} + \frac{9aabb}{\overline{a + b}^4} + \frac{27a^3b^3}{\overline{a + b}^6}} \ \&c.$$ which, in

the Cafe of an Equality of *Skill* between the Gamefters, will be reduced to this Series,

$$\frac{2}{8} \times \overline{1 + \overset{\text{III}}{\frac{3}{4}} + \overset{\text{V}}{\frac{9}{16}} + \overset{\text{VII}}{\frac{27}{64}}} \ \overset{\text{IX}}{\&c.}$$ whofe Terms are refpective-

ly correfponding to the number of Games 3, 5, 7, 9 &c. Wherefore fo many of thofe Terms ought to be taken, as that their fum being multiplied by the common Multiplicator

tor $\frac{2}{8}$ or $\frac{1}{4}$, the Product may be equal to the Fraction $\frac{1}{2}$, which Fraction denotes the equal Probability of an Events Happening or not Happening: But if two of those Terms be taken, and that their sum be Multiplied by $\frac{1}{4}$, the Product will be $\frac{7}{16}$; which being less than the Fraction $\frac{1}{2}$, it may be concluded that Five Games are too few to make it as Probable that the Play will be Ended in that number of Games as not; and that the Odds against its Ending in Five Games are 9 to 7. But if Three of those Terms be taken, then their sum being multiplied by the common Multiplicator $\frac{1}{4}$, the Product will be $\frac{37}{64}$; which exceeding the Fraction $\frac{1}{2}$, it may be concluded that Seven Games are too many; and that the Odds of the Play being Ended in Seven Games, or sooner, are 37 to 27, or 4 to 3 very nearly.

N. B. It would be needless to inquire whether Six Games might not bring the Play to an equal Probability of Ending or not Ending; it having been observed before, that in the Case of an equality of Stakes to be play'd for, it is impossible that the Play should End in an Even number of Games, if the number of Stakes be Odd; or that it should End in an Odd number of Games, if the number of Stakes be Even.

In like manner, if the Play were to continue till Four Stakes be won or lost on either side: Then taking the following Series, *viz.*

$$\frac{a^4+b^4}{\overline{a+b}|^4} \times \overline{1 + \frac{4ab}{\overline{a+b}|^2} + \frac{14aabb}{\overline{a+b}|^4} + \frac{48a^3b^3}{\overline{a+b}|^6}} \text{ &c. which, upon}$$

the supposition of an equality of Skill between the Gamesters, may be reduced to this, *viz.*

$$\frac{2}{16} \times \overset{\text{IV}}{1} + \overset{\text{VI}}{\frac{4}{4}} + \overset{\text{VIII}}{\frac{14}{16}} + \overset{\text{X}}{\frac{48}{64}} + \overset{\text{XII}}{\frac{164}{256}} \text{ &c. let so many}$$

of its Terms be tried, as will make the Product of their sum multiplied by $\frac{2}{16}$, equal to the Fraction $\frac{1}{2}$, or as near it as possible. Now Five of those Terms being tried, and their sum being multiplied by $\frac{2}{16}$, or $\frac{1}{8}$, the Product will be $\frac{1092}{2048}$, which not differing much from $\frac{1}{2}$, it may be

con-

concluded that Twelve will be very near that number of Games, which will make the Probabilities of the Plays End-ing or not Ending to be equal; the Odds for its Ending being only 1092 to 956, or 8 to 7 very nearly. But the Odds against its Ending in Ten Games, will be found to be 39 to 29, or 4 to 3 nearly.

By the same method of Process, it will be found that Five Stakes will probably be won or lost in about Seven-teen Games: It being but the Odds of 11 to 10 nearly, that the Play will not be Ended in that number of Games, and 10 to 9 nearly, that it will be Ended in Nineteen.

It will also be found that Six Stakes will probably be won or lost in about Twenty Six Games, there being but the Odds of 168 to 167 nearly, that the Play will not be End-ed in that number of Games, and 25 to 22 nearly, that it will be Ended in Twenty Eight.

If the same Method of Trial be applied to any other num-ber of Stakes, whether equal or unequal, and to any pro-portion of Skill, the number of Games required will always be found.

Yet if the number of Stakes were great, those Trials would become tedious, notwithstanding the Help that might be derived from our Second *Lemma*, whereby any number of Terms of those Series which are employed in the Solution of this Problem, may be added together. For which reason it will be convenient to make some Trials of another nature, and to see whether, from the resolution of some of the sim-plest Cases of this Problem, any Analogy can be observed between the number of Stakes given, and the number of Games which determine the equal Probability of the Plays Ending or not Ending.

Now Mr *de Monmort* having with great Sagacity disco-vered that Analogy, in the Case of an equal and Odd num-ber of Stakes, on supposition of an equality of Skill between the Gamesters, I thought the Reader would be well pleased to be acquainted with the Rule which he has given for that purpose, and which is as follows.

Let n be any Odd number of Stakes to be won or lost on either side; let also $\frac{n+1}{2}$ be made equal to p: Then the Quantity $3pp - 3p + 1$ will denote a number of Games,

wherein

wherein it will be more than an equal Chance that the Play will be Ended; thus, if the number of Stakes be Nineteen, then p will be 10, and the Quantity $3pp - 3p + 1$ will be 271, which shews that 'tis more than an equal Chance that the Play will be Ended in 271 Games.

The Author of this Rule owns that he has not been able to find another like it, for an Even number of Stakes; but I am of opinion, that tho' the same Rule, being applied to that Case, may not find the juft number of Games wherein there will be more than an equal Probability of the Plays Ending, yet it will always find a number of Games, wherein it is very near an equal Wager that the Play will be Ended. Wherefore to make the Rule as extensive as it may be, I would Chufe to exprefs it by the number of Stakes whether Even or Odd, and make it $\frac{1}{4} nn$, which differs from his own, but by the fmall Fraction $\frac{1}{4}$.

If any one has a mind to carry this fpeculation ftill farther, and to try whether fome general Rule may not be difcovered for determining, by a very near approximation, the number of Games requifite to make it a Wager of any given proportion of Odds, that the Play will be Ended in that number of Games, whether the Skill of the Gamefters be equal or unequal; let him Solve feveral Cafes of this Problem in the following manner, which I take to be as expeditious as the nature of the Problem can admit of.

Upon a Diameter equal to Unity, if fo be the Skill of the Gamefters be equal; or to the Quantity $\frac{4ab}{a+b|^2}$, if their Skill be in the proportion of a to b, let a Semicircle be defcribed, which divide into fo many equal parts as there are Stakes to be won or loft on either fide, fuppofing thofe Stakes to be equal. From the Firft, Third, Fifth, Seventh &c. Points of Divifion, beginning from one extremity of the Diameter, let Perpendiculars fall upon that Diameter, which by their concourfe with it, fhall determine the *verfed Sines* of fo many Arcs, to be taken from the other extremity thereof. Let the greateft of thofe *verfed Sines* be called m, the next lefs p, the next to it q, the next s &c. Make alfo

Q q $\qquad\qquad$ $\overline{1 - p}$

$$\frac{\overline{1-p}\times\overline{1-q}\times\overline{1-s}}{\overline{m-p}\times\overline{m-q}\times\overline{m-s}}\ \&c. = A$$

$$\frac{\overline{1-q}\times\overline{1-s}\times\overline{1-m}}{\overline{p-q}\times\overline{p-s}\times\overline{p-m}}\ \&c. = B$$

$$\frac{\overline{1-s}\times\overline{1-m}\times\overline{1-p}}{\overline{q-s}\times\overline{q-m}\times\overline{q-p}}\ \&c. = C$$

$$\frac{\overline{1-m}\times\overline{1-p}\times\overline{1-q}}{\overline{s-m}\times\overline{s-p}\times\overline{s-q}}\ \&c. = D$$

&c.

then will the Probability of the Play's not Ending in a number of Games denominated by x, be expreſt by the Quantities

$$m^{\frac{1}{2}x}A + p^{\frac{1}{2}x}B + q^{\frac{1}{2}x}C + s^{\frac{1}{2}x}D\ \&c.$$ if the number of Stakes be Even, or by the Quantities

$$m^{\frac{x-1}{2}}A + p^{\frac{x-1}{2}}B + q^{\frac{x-1}{2}}C + s^{\frac{x-1}{2}}D\ \&c.$$ if the number of Stakes be Odd.

EXAMPLE I.

LET it be required to find what Odds there is, that in 40 Games there will be Four Stakes won or loſt on either ſide.

Having divided the Semicircle into Four equal parts, according to the abovementioned directions, the Quantity m will be the *Verſed Sine* of 135 Degrees, and the Quantity p will be the *Verſed Sine* of 45 Degrees, which by the help of a Table of *Sines* will readily be found to be 0.85355 and 0.14645 reſpectively. Moreover the Quantity A being equal to $\frac{1-p}{m-p}$, and the Quantity B to $\frac{1-s}{p-s}$, will be found to be 1.2071 and — 0.2071. From whence it follows, that the Probability of the Plays not Ending in Forty Games may be expres'd by the two following Products $\overline{0.85355}^{20} \times 1.2071$ — $\overline{0.14645}^{20} \times 0.2071$, of which the Second may be entirely neglected, as being inconſiderably little in reſpect of
the

the firſt. Now the Logarithm of the firſt Product being
2.7063225, to which anſwers the number 0.05085, let that
number be ſubtracted from Unity; and the remainder being
0.94915, I conclude that the Odds of the Plays Ending in
Forty Games are as 94915 to 5085, or very near as 19
to 1.

EXAMPLE II.

LET it be required to find how many Games muſt be
play'd, to make it a Wager of 100 to 1, that Four
Stakes will be won or loſt on either ſide, in that number of
Games.

Let x be the number of Games required: Then by the
foregoing Example it will appear that we may have the
Equation $\overline{0.85355}|^{\frac{1}{2}x} \times 1.2071 = \frac{1}{100}$, in which the value
of x may eaſily be obtained by Logarithms; it being found
by one ſingle Diviſion to be about 60.

If the Stakes be unequal, the Solution will conſiſt of two
Series, in both which the Quantities m, p, q &c. will be of the
ſame value, and will be determined likewiſe by a Table of
Sines. In this Caſe the Semicircumference ought to be divided into as many equal parts as there are Units in the
number of all the Stakes: Thus, if the Stakes were Four
and Five, the Semicircumference ought to be divided into Nine
equal parts: But then it is to be obſerved that the *verſed
Sines* of thoſe Arcs, which, in the Caſe of Nine Stakes for
each Gameſter, are alternately omitted, are thoſe which, in
the Caſe of Four and Five, are to repreſent the Quantities m, p, q &c. It is to be obſerved alſo that the Quantities A, B, C, D &c. by which the Terms of the firſt Series
are to be reſpectively Multiplied, will be found to differ
from the Quantities A', B', C', D' &c. by which the Terms
of the ſecond Series are alſo to be reſpectively Multiplied;
and that both thoſe Series of Quantities may be determined
by proper Theorems contrived for that purpoſe.

Before I make an End of this Subject, I ſhall propoſe an
Inquiry to be made by thoſe who have ſufficient leiſure to
Try the foregoing Methods; which is, whether the number
of Games, wherein it will be an equal Wager that the Play
will

will be Ended, upon the suppofition of an equal number *n* of Stakes to be won or loft on either fide; as alfo of the proportion of Skill expreft by *a* and *b*, may not be determined very nearly by the following Expreffion, *viz.*

$$\overline{\frac{n\,a^n - n\,b^n}{a^n + b^n}} \times \overline{\frac{aa + ab + bb}{aa - bb}}.$$

PROBLEM XLVI.

IF A *and* B, *whofe proportion of Skill is fuppofed equal, play together till Four Stakes be won or loft on either fide; and that* C *and* D, *whofe proportion of Skill is alfo fuppofed equal, play likewife together till Five Stakes be won or loft on either fide: What is the Probability that the Play between* A *and* B *will be Ended in fewer Games than the Play between* C *and* D?

SOLUTION.

THE Probability of the Firft Play's being Ended in any number of Games before the Second, is compounded of the Probability of the Firft Play's being Ended in that number of Games, and of the Second's not being Ended with the Game immediately preceding: From whence it follows, that the Probability of the Firft Plays Ending in an Indeterminate number of Games before the Second, is the fum of all the Probabilities *ad Infinitum* of the Firft Play's Ending, Multiplied by the refpective Probabilities of the Second's not being Ended with the Game immediately preceding.

But it appears from our XXXIV*th* Problem, that the Probability of the firft Play's Ending in an Indeterminate number of Games, may be expreft by the following Series, *viz.*

$$\overset{\text{IV}}{\frac{1}{2^3}} + \overset{\text{VI}}{\frac{6}{2^5}} + \overset{\text{VIII}}{\frac{14}{2^7}} + \overset{\text{X}}{\frac{48}{2^9}} + \overset{\text{XII}}{\frac{164}{2^{11}}} + \overset{\text{XIV}}{\frac{560}{2^{13}}} \text{ \&c.}$$

It appears alfo, from our XXXIII*d* Problem, that the Probability of the Second Play's not Ending may be expreft by the following Series, *viz.*

$$\overset{\text{III}}{\frac{4}{2^2}} + \overset{\text{V}}{\frac{15}{2^4}} + \overset{\text{VII}}{\frac{55}{2^6}} + \overset{\text{IX}}{\frac{200}{2^8}} + \overset{\text{XI}}{\frac{725}{2^{10}}} + \overset{\text{XIII}}{\frac{2625}{2^{12}}} \text{ \&c.}$$

Now

Now the Corresponding Terms of thofe two Series being Multiplied together, the Products, fuppofing r equal to the Fraction $\frac{1}{16}$, will compofe the following Series, *viz.*

$$2r + 30rr + 385r^3 + 4800r^4 + 59400r^5 \text{ &c.}$$ in which Series the Index of the Relation of each Numerical Quantity to the preceding ones, may be found by the help of our Third *Lemma*: For the Index of the Relation in the Numerator of the Firft Series being $4 - 2$, and the Index of the Relation in the Numerator of the Second being $5 - 5$, which Relations are deduced from the XXXIV*th* Problem, it follows, that if in the Theorem of our Third *Lemma*, the Quantities $4, -2, 5, -5$, be refpectively fubftituted in the room of the Quantities m, n, p, q, the Index of the Relation in the Third Series will be found to be $20 - 110 + 200 - 100$; wherefore all the Terms of this Series may be fummed up by the Third Theorem of our Second *Lemma*, fubftituting the Quantities 20, 110, 200, 100 in the room of the Quantities m, n, p, q, therein employed; fubftituting alfo the Terms $2r, 30rr, 385r^3, 4800r^4$ in the room of the Quantities A, B, C, D: For after thofe Subftitutions, the fum of the Third Series will be found to be $\frac{2r - 10rr + 5r^3}{1 - 20r + 110rr - 200r^3 + 100r^4}$, which is reduced to $\frac{476}{723}$ by changing the Quantity r into its value $\frac{1}{16}$. Now fubtracting the Fraction $\frac{476}{723}$ from Unity, the remainder will be the Fraction $\frac{247}{723}$, the Numerators of which two Fractions exprefs the Odds of the Firft Plays Ending before the Second, which confequently will be as 476 to 247, or 27 to 14 nearly.

If in the foregoing Problem, the Skill of the Gamefters had been in any proportion of inequality, the Problem might have been Solved with the fame eafe.

When in a Problem of this nature the number of Stakes to be loft by either A or B, does not exceed the number Three, the Problem may be always readily Solved without the ufe of the Theorem inferted in our Third *Lemma*; tho' the number of Stakes between C or D be never fo great. For which reafon, if any one has the curiofity to try, if from the Solution of feveral Cafes of this Problem, fome Rule may not be difcovered for Solving the fame generally; it will be

R r

con-

convenient he fhould compare together the different Solutions, which may refult from the fuppofition that the Stakes to be loft by either *A* or *B* are Two or Three; and then the Cafe of the foregoing Problem may alfo be compared with all the reft: Yet as thefe Trials might not perhaps be fufficient to difcover any Analogy between thofe Solutions, I have thought fit to add a new Theorem in this place, whereby Four Cafes more of this Problem may be Solved, *viz.* When the number of Stakes to be loft by *A* or *B*, and by *C* or *D*, are 4 and 6, 4 and 7, 5 and 6, 5 and 7: The Theorem being as follows.

If there be a Series of Terms whofe Relation is expreffed by the Index $l + m + n$, and there be likewife another Series of Terms whofe Relation is expreffed by the Index $p + q$; and the Correfponding Terms of thofe two Series be Multiplied together: Then the Index of the Relation in the Third Series, refulting from the Multiplication of their correfponding Terms, will be expreffed by the Quantities.

$$lp + \begin{matrix} + 2mq + lmpq + 2lnqq \\ llq + np^3 - mmqq - mnpqq + nnq^2. \\ + mpp + 3npq + lnppq \end{matrix}$$

It is to be obferved, that altho' thefe forts of Theorems might be applicable to the finding of the Relation of thofe Terms, which are the Products of the correfponding Terms of two different Series, both of which confift of Terms whofe laft Differences are equal to nothing; yet there will be no necefity to ufe them for that purpofe, that Relation being to be found much fhorter, as follows.

Let e and f denote the rank of thofe Differences which are refpectively equal to nothing in each Series; then the Quantity $e + f - 1$ will denote the rank of that Difference which is equal to nothing, in the Series refulting from the Multiplication of the correfponding Terms of the other two; and confequently the Relation of the Terms of this New Series will eafily be obtained by our firft *Lemma.*

AFter having given the Solution of several sorts of Problems, each of them containing some degree of Difficulty not to be met with in any of the rest; and having thereby laid a sufficient foundation for solving the most intricate cases that may occurr in this Subject of Chances, it might almost seem superfluous to add any thing to this Tract: Yet considering that a Variety of Examples is the properest means of making Rules easy and familiar; and designing to be as useful as possible to those of my Readers, who perhaps may not be so well versed in Algebraical Calculations, I have chose to fill up the remaining Pages of this Book, with some easy Problems relating to the Games which are most in use; such as HAZARD, WHISK, PIQUET, &c. and to enlarge a little more upon the Doctrine of Combinations.

PROBLEM XLVII.

TO find at HAZARD the *Advantage* of the Setter upon all *Suppositions* of Main and Chance.

SOLUTION.

LET the whole Money Play'd for be considered as a common Stake, upon which both the Setter and Caster have their several Expectations; then let those Expectations be determined in the following manner.

First, Let it be supposed that the Main is *vii*; then if the Chance of the Caster be *vi* or *viii*, it is plain that the Setter having Six Chances to win and Five to lose, his Expectation will be $\frac{6}{11}$ of the Stake: But there being Ten Chances

out

out of Thirty-fix for the Chance to be *vi* or *viii*, it follows, that the Expectation of the Setter, refulting from the Probability of the Chance being *vi* or *viii*, will be $\frac{10}{36}$ multiplyed by $\frac{6}{11}$, or $\frac{60}{11}$ to be divided by 36.

Secondly, If the Main being *vii*, the Chance fhould Happen to be *v* or *ix*; then the Setter having Six Chances to win and Four to lofe, his Expectation will be $\frac{6}{10}$ or $\frac{3}{5}$ of the Stake: But there being Eight Chances in Thirty-fix for the Chance to be *v* or *ix*, it follows, that the Expectation of the Setter, refulting from the Probability of that Chance, will be $\frac{8}{36}$ multiplied by $\frac{3}{5}$, or $\frac{24}{5}$ to be divided by 36.

Thirdly, If the Main being *vii*, the Chance fhould Happen to be *iv* or *x*; then the Setter having Six Chances to win and Three to lofe, his Expectation will be $\frac{6}{9}$ or $\frac{2}{3}$ of the Stake: But there being Six Chances out of Thirty-fix for the Chance to be *iv* or *x*, it follows, that the Expectation of the Setter, refulting from the Probability of that Chance, will be $\frac{6}{36}$ multiplied by $\frac{2}{3}$, or 4 divided by 36.

Fourthly, If the Main being *vii*, the Cafter fhould Happen to throw *ii*, *iii*, or *xii*; then the Expectation of the Setter will be the whole Stake, for which there being Four Chances in Thirty-fix, it follows, that the Expectation of the Setter, refulting from the Probability of thofe Cafes, will be $\frac{4}{36}$ of the Stake, or 4 divided by 36.

Lastly, If the Main being *vii*, the Cafter fhould Happen to throw *vii* or *xi*, the Setter lofes his Expectation.

From the Solution of the foregoing particular Cafes it follows, that the Main being *vii*, the Expectation of the Setter will be expreft by the following Quantities, *viz.* $\dfrac{\frac{60}{11} + \frac{24}{5} + \frac{4}{1} + \frac{4}{1}}{36}$

which may be reduced to $\frac{251}{495}$. Now this Fraction being fubtracted from Unity, to which the whole Stake is fuppofed equal, there will remain the Expectation of the Cafter *viz.* $\frac{244}{495}$.

But the Probabilities of winning being always proportional to the Expectations, on fuppofition of the Stake being fixt, it follows, that the Probabilities of winning for the Setter
and

and Cafter are refpectively Proportional to the two numbers
251 and 244, which properly denote the Odds of winning.

Now, if we fuppofe each Stake to be 1, or the whole Stake
to be 2, the Gain of the Setter will be exprest by the Fraction
$\frac{7}{495}$, it being the Difference of the Odds divided by their Sum,
which fuppofing each Stake to be a Guinea, will be about
3 *d* : 2 $\frac{1}{2}$ *f.*

By the fame Method of Procefs, it will be found that the
Main being *vi* or *viii*, the Gain of the Setter will be $\frac{167}{7128}$,
which is about 6 *d* : $\frac{1}{6}$ *f* in a Guinea.

It will alfo be found that the Main being *v* or *ix*, the
Gain of the Setter will be $\frac{43}{2835}$, which is about 4 *d* : 2 $\frac{1}{9}$ *f*
in a Guinea.

Coroll. 1. If each particular Gain made by the Setter, in
the Cafe of any Main, be refpectively Multiplied by the
number of Chances there are for that Main to come up,
and the Sum of the Products be divided by the number of
all thofe Chances, the Quotient will exprefs the Gain of the
Setter before a Main is thrown: from whence it follows, that
the Gain of the Setter, if he be refolved to fet upon the firft
Main, may be eftimated to be $\frac{344}{2835}$ + $\frac{1670}{7128}$ + $\frac{42}{495}$ to be
divided by 24; which being reduced will be $\frac{9}{109}$ very
nearly, or about 4 *d* : 2 $\frac{1}{10}$ *f.*

Coroll. 2. The Probability of no Main is to the Probability of a Main, as 109 + 2 to 109 − 2, or as 111 to 107.

Coroll. 3. The Lofs of the Cafter's hand, if each throw
be for a Guinea, and he confine himfelf to hold it as long as
he wins, will be $\frac{4}{111}$ or about 9 *d.* in all, the Demonftration
of which may be deduced from our XXIV*th* Problem.

PROBLEM XLVIII.

IF *Four Gamefters play at* WHISK; *What are the Odds that
any two of the Partners that are pitch'd upon, have not the
Four Honours?*

S f   SOLU-

SOLUTION.

First, suppose those two Partners to have the Deal, and the last Card which is turn'd up to be an Honour.

From the suppofition of these two Cafes, we are only to find what Probability the Dealers have of taking Three fet Cards in Twenty five, out of a Stock containing Fifty one. To refolve this the fhorteft way, recourfe muft be had to the Theorem given in the *Corollary* of our XX*th* Problem, in which making the Quantities *n, c, d, p, a,* refpectively equal to the numbers 51, 25, 26, 3, 3, the Probability required will be found to be $\frac{25 \times 24 \times 23}{51 \times 50 \times 49}$ or $\frac{92}{833}$.

Secondly, If the Card which is turn'd up be not an Honour, then we are to find what Probability the Dealers have, of taking Four given Cards in Twenty five out of a Stock containing Fifty one, which by the aforefaid Theorem will be found to be $\frac{25 \times 24 \times 23 \times 22}{51 \times 50 \times 49 \times 48}$ or $\frac{253}{4998}$.

But the Probability of taking the Four Honours being to be eftimated before the laft Card is turn'd up; and there being Sixteen Chances in Fifty two, or Four in Thirteen for an Honour to turn up; and Nine in Thirteen againft it; it follows, that the Fraction expreffing the Probability of the Firft Cafe ought to be Multiplied by 4; that the Fraction expreffing the Probability of the Second ought to be Multiplied by 9; and that the fum of thofe Products ought to be divided by 13; which being done, the Quotient $\frac{115}{1666}$, or $\frac{2}{29}$ nearly, will exprefs the Probability required.

Corollary, By the help of the abovecited Theorem, the following Conclufions may eafily be verified.

It is 27 to 2 nearly that the two Dealers have not the Four Honours.

It is 23 to 1 nearly that the two Eldeft have not the Four Honours.

It is 8 to 1 nearly that neither one Side nor the other have the Four Honours.

It

It is 13 to 7 nearly that the two Dealers do not reckon Honours.

It is 20 to 7 nearly that the two Eldeſt do not reckon Honours.

It is 25 to 16 nearly that either one Side or the other do reckon Honours, or that the Honours are not equally divided.

PROBLEM XLIX.

Of RAFFLING.

IF *any number of Gameſters A, B, C, D &c. Play at* Raffles: *What is the Probability that the firſt of them having got his Chance wins the Money of the Play?*

SOLUTION.

IN order to Solve this Problem, it is neceſſary to have a Table ready compos'd, of all the Chances which there are in three Raffles, which Table is the following. Wherein

The firſt Column contains the number of Points which are ſuppoſed to have been thrown by *A* in three Raffles.

The ſecond Column contains the number of Chances which *A* has to win if his Points be above *xxxi,* or the number of Chances he has to loſe if they be either *xxxi* or below it.

The third Column contains the number of Chances which *A* has to loſe, if his Points be above *xxxi,* or to win if they be either *xxxi* or below it.

The Fourth Column contains the number of Chances which he has for an equality of Chance.

The Conſtruction of this Table eaſily flows from the conſideration of the number of Chances which there are in a ſingle Raffle; whereof *xviii* or *iii,* have 1 Chance; *xvii* or *iv,* 3 Chances; *xvi* or *v,* 6 Chances; *xv* or *vi,* 4 Chances; *xiv* or *vii,* 9 Chances; *xiii* or *viii,* 9 Chances ; *xii* or *ix,* 7 Chances ; *xi* or *x,* 9 Chances; which number of Chances being duly Combined will afford all the Chances of Three Raffles.

A T A

A TABLE of all the CHANCES which are in three *Raffles*.

Points		Chances to win or lose.	Chances to win or lose.	Equality of Chance.
liv	*ix*	884735	0	1
liii	*x*	884726	1	9
lii	*xi*	884681	10	45
li	*xii*	884534	55	147
l	*xiii*	884165	202	369
xlxix	*xiv*	883400	571	765
xlviii	*xv*	881954	1336	1446
xlvii	*xvi*	879470	2782	2484
xlvi	*xvii*	875501	5266	3969
xlv	*xviii*	869632	9235	5869
xliv	*xix*	861199	15104	8433
xliii or *xx*	849706	23537	11493
xlii	*xxi*	834679	35030	15027
xli	*xxii*	815392	50057	19287
xl	*xxiii*	791506	69344	23886
xxxix	*xxiv*	762838	93230	28668
xxxviii	*xxv*	728971	121898	33867
xxxvii	*xxvi*	690100	155765	38871
xxxvi	*xxvii*	646929	194636	43171
xxxv	*xxviii*	599472	237807	47457
xxxiv	*xxix*	548865	285264	50607
xxxiii	*xxx*	496314	335871	52551
xxxii	*xxxi*	442368	388422	53946

Sum 442368
442368
884736

This

This being once fuppofed, let it be required to find the Probability which *A* has of winning, when the number of his Points being *xl,* there is but one Gamefter *B* befides himfelf.

Take the number 791506, which in the fecond Column ftands over againft the number *xl,* to be found in the Firft. Take alfo one half of the number which in the Fourth Column ftands over againft the faid Number *xl,* which half is 11943. Let thefe two Numbers *viz.* 791506 and 11943 be added together, and their Sum 803449 being divided by 884736, which is the Number of all the Chances, the Quotient, *viz.* $\frac{803449}{884736}$ will exprefs the Probability required.

Now this Fraction being Subtracted from Unity, and the remainder being $\frac{81287}{884736}$, it follows that the Numerators of thefe two Fractions, *viz.* 803449 and 81287 do exprefs the Odds of winning, which may be reduc'd to 89 and 9 nearly.

But if the Number of Points which *A* has thrown for his Chance being *xl* as above, there be two other Gamefters *B* and *C* befides himfelf, the Probability which he has of winning will be found thus.

Take the Square of the Number fet down over againft *xl* in the fecond Column, which Square is 626481748036. Take alfo the Product of that Number by the Number fet down over againft *xl* in the Fourth Column, which Product is 18905912316. Laftly, take the third part of the Square of the Number fet down in the Third Column, which third part will be 190180332, and let all thofe numbers be added together: Then their Sum being divided by the Square of the whole Number of Chances, *viz.* by 782757789696, the Quotient $\frac{645577840684}{782757789696}$ will exprefs the Probability required; from whence it may be concluded that the Odds of winning are nearly as 33 to 7.

N. B. If fome of the laft figures in the Numbers of the foregoing Table be neglected, the Operation will be fhortned, and a fufficient Approximation obtain'd by help of the remaining Figures.

From what we have faid it follows, that *A* having *xl* for the number of his Points, has lefs advantage when he Plays against One than when he plays against Two: For supposing each Man's Stake be a Guinea, he has in the first Case 89 Chances for winning 1, and 9 Chances for losing 1:

From whence it follows that his Gain is $\frac{80-9}{98}$ or $\frac{80}{98}$ which is about 17 *fb.* 6 *d.*

But in the second Case, supposing also each Man's Stake to be a Guinea, he has 33 Chances for winning 2, and 7 Chances for losing 1:

Whence it appears, that his Gain in this Case is $\frac{2\times33-7}{40}$ or $\frac{59}{40}$ which is about 1 *l.* — 12 *fb.* — 8 *d.* But Note, that it is not to be concluded from this single Instance, that the Gain of *A* will always increase with the number of Gamesters.

If the number of Gamesters be never so many, let *p* be their number, let *a* be the number of Chances which *A* has for winning when he has thrown his Chance, let *m* be the number of Chances which there are for an Equality of Chance between *A* and any of the other Gamesters; Lastly, let the whole number of Chances be denoted by *S*: Then the Probability which *A* has of winning will be expressed by the following Series.

$$\frac{a^{p-1} + \frac{p-1}{2} m\, a^{p-2} + \frac{p-1}{2} \times \frac{p-2}{3} mm\, a^{p-3} + \frac{p-1}{2} \times \frac{p-2}{3} \times \frac{p-3}{4} m^3 a^{p-4} \&c.}{S^{p-1}}$$

which Series is composed of the Terms of the Binomial $\overline{a+m}^{\,p-1}$ reduced into a Series, all its Terms being divided by 1, 2, 3, 4, 5 &c. respectively.

The foregoing Theorem may be useful, not only for solving any Case of the present Problem, but also an infinite Variety of other Cases, in those Games wherein there is no Advantage in the order of Play: And the Application of it to Numbers will be found easy, to those who understand how to use Logarithms.

PROBLEM L.

TO *find what Probability there is, that any Number of Cards of each Suit may be contained in a given number of them taken out of a given Stock.*

SOLUTION.

First, Find the whole number of Chances there are for taking the given number of Cards out of the given Stock.

Secondly, Find all the particular Chances there are for taking each given number of Cards of each Suit out of the whole number of Cards belonging to that Suit.

Thirdly, Multiply all those particular Chances together; then divide the Product by the whole number of Chances, and the Quotient will express the Probability required.

Thus, If it be proposed to find the Probability of taking Four Hearts, Three Diamonds, Two Spades and One Club, in Ten Cards taken out of a Stock containing Thirty-two.

Find the whole number of Chances for taking ten Cards out of a Stock containing two and Thirty; which is properly Combining two and Thirty Cards Ten and Ten. To do this, write down all the Numbers from 32 inclusively to 22 exclusively, so as to have as many Terms as there are Cards to be Combined; then write under each of them respectively all the numbers from One to Ten inclusively, thus,

$$\frac{32 \times 31 \times 30 \times 29 \times 28 \times 27 \times 26 \times 25 \times 24 \times 23}{1 \times 2 \times 3 \times 4 \times 5 \times 6 \times 7 \times 8 \times 9 \times 10}$$

Let all the numbers of the upper Row be Multiplied together; let also all the numbers of the lower Row be Multiplied together, then the first Product being divided by the second, the Quotient will express the whole number of Chances required, which will be 64512240.

By the like Operation the number of Chances for taking Four Hearts out of Eight, will be found to be

$$\frac{8 \times 7 \times 6 \times 5}{1 \times 2 \times 3 \times 4} = 70.$$

The

The number of Chances for taking Three Diamonds out of Eight will alfo be found to be $\frac{8 \times 7 \times 6}{1 \times 2 \times 3} = 56$.

The number of Chances for taking Two Spades out of Eight will in the fame manner be found to be $\frac{8 \times 7}{1 \times 2} = 28$.

Laftly, The number of Chances for taking One Club out of Eight will be found to be $\frac{8}{1} = 8$.

Wherefore, Multiplying all thefe particular Chances together *viz.* 70, 56, 28, 8, the Product will be 878080; which being Divided by the whole number of Chances, the Quotient $\frac{878080}{64512240}$, or $\frac{2}{101}$ nearly, will exprefs the Probability required: From whence it follows, that the Odds againft taking Four Hearts, Three Diamonds, Two Spades and One Club in Ten Cards, are very near 99 to 2.

It is to be obferved that the Operations whereby the Number of Chances is determined, may always be contracted, except in the fingle cafe of taking one Card only of a given Suit. Thus, If it were propofed to fhorten the Fraction

$$\frac{32 \times 31 \times 30 \times 29 \times 28 \times 27 \times 26 \times 25 \times 24 \times 23}{1 \times 2 \times 3 \times 4 \times 5 \times 6 \times 7 \times 8 \times 9 \times 10},$$ which determines the number of all the Chances belonging to the foregoing Problem: Let it be confidered whether the Product of any two or more Terms of the Denominator, being Multiplied together, be equal to any one of the Terms of the Numerator; if fo, all thofe Terms may be expunged out of both Denominator and Numerator. Thus the Product of the three Numbers 2, 3, 4, which are in the Denominator, being equal to the Number 24, which is in the the Numerator, it follows, that the three Numbers 2, 3, 4 may be expunged out of the Denominator, and at the fame time the Number 24 out of the Numerator. For the fame reafon the Numbers 5 and 6 may be expunged out of the Denominator, and the Number 30 out of the Numerator, which will reduce the Fraction to be

$$\frac{32 \times 31 \times 30 \times 29 \times 28 \times 27 \times 26 \times 25 \times 24 \times 23}{1 \times 2 \times 3 \times 4 \times 5 \times 6 \times 7 \times 8 \times 9 \times 10}$$

It ought likewife to be confidered whether there be any of the remaining Numbers in the Denominator that Divide exactly any of the remaining Numbers of the Numerator.

If

If so, those Numbers are to be expunged out of the Denominator and Numerator, but the respective Quotients of the Terms of the Numerator divided by those of the Denominator, are to be substituted in the room of those Terms of the Numerator. Thus the Terms 7, 8, and 9 of the Denominator dividing exactly the Numbers 28, 32 and 27 of the Numerator, and the Quotients being 4, 4 and 3 respectively, all the Numbers 7, 8, 9, 28, 32, 27 ought to be expunged, and the Quotients 4, 4, and 3 substituted in the room of 28, 32, 27 respectively, in the following manner;

$$\frac{4 \qquad\qquad\qquad 4 \quad 3}{32 \times 31 \times 30 \times 29 \times 28 \times 27 \times 26 \times 25 \times 24 \times 23}{1 \times 2 \times 3 \times 4 \times 5 \times 6 \times 7 \times 8 \times 9 \times 10}$$

It ought also to be considered, whether the remaining Terms of the Denominator have any common Divisor with any of the remaining Terms of the Numerator; if so, dividing those Terms by their common Divisors, the respective Quotients ought to be substituted in the room of the Terms of the Numerator. Thus, the only remaining Term in the Denominator, besides Unity, being 10, which has a common Divisor with one of the remaining Terms of the Numerator, *viz.* 25, and that common Divisor being 5, let 10 and 25 be respectively divided by the common Divisor 5, and let the respective Quotients 2 and 5 be substituted in the room of them, and the Fraction will be reduced to the following, *viz.*

$$\frac{4 \qquad\qquad\qquad 4 \quad 3 \qquad\quad 5}{32 \times 31 \times 30 \times 29 \times 28 \times 27 \times 26 \times 25 \times 24 \times 23}{1 \quad 2 \quad 3 \quad 4 \quad 5 \quad 6 \quad 7 \quad 8 \quad 9 \quad 10}{2}$$

Lastly, Let the remaining Number 2 in the Denominator divide any of the Numbers of the Numerator which are divisible by it, such as 26, and let those two Numbers be expunged; but let the Quotient of 26 by 2, *viz.* 13, be substituted in the room of 26: And then the Fraction, neglecting unity, which is the only Term remaining, may be reduced to 4 × 31 × 29 × 4 × 3 × 13 × 5 × 23, the Product of which Numbers is 64512240, as we have found it before.

The foregoing Solution being well understood, it will be easy to enlarge the Problem, and to find the Probability of

U u taking

taking at leaſt Four Hearts, Three Diamonds, Two Spades, and One Club, in Eleven Cards; the finding of which depends upon the four following Caſes, *viz.* taking.

    5 Hearts, 3 Diamonds, 2 Spades, 1 Club,
    4 Hearts, 4 Diamonds, 2 Spades, 1 Club,
    4 Hearts, 3 Diamonds, 3 Spades, 1 Club,
    4 Hearts, 3 Diamonds, 2 Spades, 2 Clubs.

Now the number of Chances for the Firſt Caſe will be found to be 702464, for the Second 1097600, for the Third 1756160, for the Fourth 3073280; which Chances being added together, and their ſum divided by the whole number of Chances for taking Eleven Cards out of Thirty two, the Quotient will be $\frac{6628504}{129024480}$, which may be reduced to $\frac{1}{97}$ nearly.

From whence it may be concluded, that the Odds againſt the taking of Four Hearts, Three Diamonds, Two Spades, and One Club, in Eleven Cards, that is, ſo many at leaſt of every ſort, is about 92 to 5.

And by the ſame Method it would be eaſy to ſolve any other Caſe of the like nature, let the number of Cards be what it will.

PROBLEM LI.

TO find at PIQUET *the Probability which the Dealer has for taking One Ace or more in Three Cards, having none in his Hands.*

SOLUTION.

FRom the number of all the Cards, which are Thirty two, ſubtracting Twelve which are in the Dealers Hands, there remains Twenty, among which are the Four Aces.

From whence it follows, that the number of all the Chances for taking any three Cards in the Bottom, are the number of Combinations which Twenty Cards may afford, being taken Three and Three; which, by the Rule given in the preceding Problem, will be found to be $\frac{20 \times 19 \times 18}{1 \times 2 \times 3}$ or 1140.

The

The number of all the Chances being thus obtained, find the number of Chances for taking one Ace precisely, with two other Cards; find next the number of Chances for taking Two Aces precisely with any other Card; Lastly, find the number of Chances for taking Three Aces: Then these Chances being added together, and their sum divided by the whole number of Chances, the Quotient will express the Probability required.

But by the Directions given in the preceding Problem, it appears, that the number of Chances for taking One Ace precisely are $\frac{4}{1}$ or 4; and that the number of Chances for taking any two other Cards are $\frac{16 \times 15}{1 \times 2}$ or 120 : From whence it follows, that the number of Chances for taking One Ace precisely with any two other Cards, is equal to 4 x 120 or 480.

In like manner it appears, that the number of Chances for taking Two Aces precisely is equal to $\frac{4 \times 3}{1 \times 2}$ or 6, and that the number of Chances for taking any other Card is $\frac{16}{1}$ or 16; from whence it follows, that the number of Chances for taking Two Aces precisely with any other Card is 6 x 16 or 96.

Lastly, It appears that the Number of Chances for taking Three Aces is equal to $\frac{4 \times 3 \times 2}{1 \times 2 \times 3}$ or 4.

Wherefore the Probability required will be found to be $\frac{480 + 96 + 4}{1140}$ or $\frac{580}{1140}$; which Fraction being subtracted from Unity, the remainder, *viz.* $\frac{560}{1140}$ will express the Probability of not taking an Ace in Three Cards: From whence it follows, that it is 580 to 560, or 29 to 28, that the Dealer takes One Ace or more in three Cards.

The preceding Solution may be very much contracted, by inquiring at first what the Probability is of not taking an Ace in Three Cards, which may be done thus:

The number of Cards in which the Four Aces are contained being Twenty, and consequently the number of Cards out of which the Four Aces are excluded being Sixteen, it follows, that the number of Chances which there are for the taking Three Cards, among which no Ace shall be found, is is the number of Combinations which Sixteen Cards may afford, being taken Three and Three; which number of Combinations by the preceding Problem will be found to be $\frac{16 \times 15 \times 14}{1 \times 2 \times 3}$ or 560.

But the number of all the Chances which there are for taking any Three Cards in Twenty, has been found to be 1140; from whence it follows, that the Probability of not taking an Ace in Three Cards is $\frac{560}{1140}$; and consequently that the Probability of taking One or more Aces in Three Cards is $\frac{580}{1140}$: The same as before.

In the like manner, if we would find the Probability which the Eldest has of taking One Ace or more in his Five Cards, he having none in his Hands; the severall Chances may be calculated as follows.

First, The number of Chances for taking One Ace and Four other Cards will be found to be 7280.

Secondly, The number of Chances for taking Two Aces and Three other Cards will be found to be 3360.

Thirdly, The number of Chances for taking Three Aces and two other Cards will be found to be 480.

Fourthly, The number of Chances for taking Four Aces and any other Card will be found to be 16.

Lastly, The number of Chances for taking any Five Cards will be found to be 15504.

Let the sum of all the particular Chances, *viz.* 7280 + 3360 + 480 + 16 or 11136, be divided by the sum of all the Chances, *viz.* by 15504, and the Quotient $\frac{11136}{15504}$ will express the Probability required.

Now the foregoing Fraction being subtracted from Unity, the remainder, *viz.* $\frac{4368}{15504}$ will express the Probability of not taking an Ace in Five Cards; wherefore the Odds of taking an Ace in Five Cards are 11136 to 4368, or 5 to 2 nearly.

But if the Probability of not taking an Ace in Five Cards be at first inquired into, the Work will be very much shortened; for it will be found to be $\frac{16 \times 15 \times 14 \times 13 \times 12}{1 \times 2 \times 3 \times 4 \times 5}$ or 4368, to be divided by the whole number of Chances, *viz.* by 15504, which makes it as before, equal to $\frac{4368}{15504}$:

But

But suppose it were required to find the Probability which the Eldeft has of taking an Ace and a King in Five Cards, he having none in his Hands: Let the following Chances be found,   *Viz.*

1 | For One Ace, One King and Three other Cards.
2 | For One Ace, Two Kings and Two other Cards.
3 | For One Ace, Three Kings and any other Card.
4 | For One Ace and Four Kings.
5 | For Two Aces, One King and Two other Cards.
6 | For Two Aces, Two Kings and any other Card.
7 | For Two Aces and Three Kings.
8 | For Three Aces, One King and any other Card.
9 | For Three Aces and Two Kings.
10 | For Four Aces and One King.
11 | For taking any Five Cards in Twenty.

Among thefe Cafes, there being four Pairs that are alike, *viz.* the Second and Fifth, the Third and Eighth, the Fourth and Tenth, the Seventh and Ninth; it follows, that there are only Seven Cafes to be Calculated, whereof the Firft, Sixth and Eleventh, are to be taken fingly; but the Second, Third, Fourth and Seventh, to be doubled. Now the Operation is as follows.

The *Firft* Cafe has $\frac{4}{1} \times \frac{4}{1} \times \frac{12 \times 11 \times 10}{1 \times 2 \times 3}$ or 3520 Chances.

The *Second,* $\frac{4}{1} \times \frac{4 \times 3}{1 \times 2} \times \frac{12 \times 11}{1 \times 2}$ or 1584, the double of which is 3168 Chances.

The *Third,* $\frac{4}{1} \times \frac{4 \times 3 \times 2}{1 \times 2 \times 3} \times \frac{12}{1}$ or 192, the double of which is 384 Chances.

The *Fourth,* $\frac{4}{1} \times \frac{4 \times 3 \times 2 \times 1}{1 \times 2 \times 3 \times 4}$ or 4, the double of which is 8 Chances.

The *Sixth,* $\frac{4 \times 3}{1 \times 2} \times \frac{4 \times 3}{1 \times 2} \times \frac{12}{1}$ or 432 Chances.

The *Seventh,* $\frac{4 \times 3}{1 \times 2} \times \frac{4 \times 3 \times 2}{1 \times 2 \times 3}$ or 24, the double of which is 48 Chances.

The *Eleventh,* $\frac{20 \times 19 \times 18 \times 17 \times 16}{1 \times 2 \times 3 \times 4 \times 5}$ or 15504, being the number of all the Chances for taking any Five Cards out of Twenty.

X x   From

From whence it follows, that the Probability whicli the Eldeſt has for taking an Ace and a King in Five Cards, he having none in his Hands, will be expreſt by the Fraction

$$\frac{3520 + 3168 + 384 + 8 + 432 + 48}{15504} \quad \text{or} \quad \frac{7560}{15504}.$$

Let this Fraction be ſubtracted from Unity, and the remainder being $\frac{7944}{15504}$, the Numerators of theſe two Fractions, *viz.* 7560 and 7944, will expreſs the proportion of Probability that there is, of taking or not taking an Ace and a King in Five Cards; which two numbers may be reduced nearly to the proportion of 20 to 21.

By the ſame Method of Proceſs, any Caſe relating to WHISK might be Calculated, tho' not ſo expeditiouſly, as by the Method explained in the *Corollary* of our XX*th* Problem: For which reaſon the Reader is deſired to have recourſe to the Method therein explained, when any other Caſe of the like nature happens to be propoſed.

PROBLEM LII.

TO *find the Probability of taking any number of Suits, in a given number of Cards taken out of a given Stock; without ſpecifying what number of Cards of each Suit ſhall be taken.*

SOLUTION.

Suppoſe the number of Cards to be taken out of the given Stock to be Eight, the number of Suits to be Four, and the number of Cards in the Stock to be Thirty-two.

Let all the Variations that may happen, in taking One Card at leaſt of each Suit, be written down in order, as follows,

1,	1,	1,	5,
1,	1,	2,	4,
1,	1,	3,	3,
1,	2,	2,	3,
2,	2,	2,	2.

Then

Then suppoſing any particular Suits to be appropriated at pleaſure to the Numbers belonging to the Firſt Caſe, as if it were required, for Inſtance, to take One Heart, One Diamond, One Spade and Five Clubs; let the Probability of the ſame be inquired into, which, by our L_{tb} Problem, will be found to be $\frac{28672}{10518300}$; but the Problem not requiring the Suits to be confined to any number of Cards of each Sort, it follows, that this Probability ought to be increaſed in proportion to the number of Permutations, or Changes of Order, which Four Things may undergo, whereof Three are alike. Now this number of Permutations is Four, and conſequently the Probability of the Firſt Caſe, that is, of taking Three Cards of three different Suits, and five Cards of a Fourth Suit, in Eight Cards, will be the Fraction $\frac{28672}{10518300}$ multiplied by 4, or $\frac{114688}{10518300}$.

In the ſame manner the Probability of the Second Caſe, ſuppoſing it were confined to One Heart, One Diamond, Two Spades, and Four Clubs, would be found to be $\frac{125440}{10518300}$; which being multiplied by 12, *viz.* by the number of Permutations which Four Sorts may undergo, whereof Two are alike, and the other Two differing, it will follow, that the Probability of the Second Caſe, taken without any reſtriction, will be expreſſed by the Fraction $\frac{1505280}{10518300}$.

The Probability of the Third Caſe will likewiſe be found to be $\frac{1204224}{10518300}$.

The Probability of the Fourth will be found to be $\frac{4214784}{10518300}$.

Laſtly, The Probability of the Fifth will be found to be $\frac{614656}{10518300}$. Theſe Fractions being added together, their ſum, *viz.* $\frac{7653632}{10518300}$ will expreſs the Probability of taking the Four Suits in Eight Cards.

Let this laſt Fraction be ſubtracted from Unity, and the remainder being $\frac{2864668}{10518300}$, it follows that 'tis the Odds of 7653632 to 2864668, or 8 to 3 nearly, that the Four Suits may be taken in Eight Cards, out of a Stock containing Thirty-two.

The only difficulty remaining in this matter, is the finding readily the number of Permutations which any number

of:

of Things may undergo, when either they be all different, or when some of them be alike. The Solution of which may be deduced from what we have said in the *Corollary* of our XVII*th* Problem, and may be explained as follows, in words at length.

Let all the numbers that are from Unity to that number which expresses how many Things are to be Permuted, be written down in order; Multiply all those Numbers together, and the Product of them all will express the number of their Permutations, if they be all different. Thus the number of Permutations which Ten things are capable of, is the Product of all the Numbers $1 \times 2 \times 3 \times 4 \times 5 \times 6 \times 7 \times 8 \times 9 \times 10$, which is equal to 3628800.

But if some of them be alike, as suppose Four of One sort, Three of another, Two of a Third, and One of a Fourth, write down as before all the Numbers $1 \times 2 \times 3 \times 4 \times 5 \times 6 \times 7 \times 8 \times 9 \times 10$; then write under them as many of those Numbers as there are Things of the First sort that are alike, which in this Case being Four, write the Numbers $1 \times 2 \times 3 \times 4$, beginning at Unity, and following in order. Write also as many of those Numbers as there are Things of the second sort that are alike, *viz.* $1 \times 2 \times 3$, still beginning at Unity. In the same manner write as many more as there are Things of the Third sort that are alike, *viz.* 1×2; and so on: Which being represented by the Fraction

$$\frac{1 \times 2 \times 3 \times 4 \times 5 \times 6 \times 7 \times 8 \times 9 \times 10}{1 \times 2 \times 3 \times 4 \times 1 \times 2 \times 3 \times 1 \times 2 \times 1},$$

let all the numbers of the upper Row be Multiplied together, let also all the numbers of the lower Row be Multiplied together, and the First Product being divided by the Second, the Quotient 12600 will express the number of Permutations required.

By this Method of Permutations, the Probability of throwing any determinate number of Faces of the like sort, with any given number of Dice, may easily be found. Thus, suppose it were required to find the Probability of throwing an Ace, a Two, a Three, a Four, a Five, and a Six with six Dice. It is plain that there are as many Chances for doing it, as there are Changes or Permutations in the
Order

Order or Place of fix different Things, fuppofe of the Six Letters *a*, *b*, *c*, *d*, *e*, *f*, which by the Rule above given would be 720, *viz.* the Product of the numbers 1, 2, 3, 4, 5, 6: For tho' the Dice are not confidered as changing their Places, or as affording any Variation upon the fcore of the different Situation they may have in Refpect to one another, being thrown upon a Table; yet they ought to be confidered as changing their Faces, which is equivalent to their changing of Place. Now the number of all the Chances upon Six Dice, being the number 6 Multiplied into it felf, as many times wanting one as there are Dice, *viz.* $6 \times 6 \times 6 \times 6 \times 6 \times 6$ or 46656, it follows, that the Probability required will be expreft by the Fraction $\frac{720}{46656}$, and confequently, that the Odds againft throwing the Faces undertaken, will be 46656 — 720 to 720, or 64 to 1 nearly.

In the fame manner fuppofe it were required to find the Probability of throwing One Ace, Two Two's and Three Three's with Six Dice. The number of Chances for the doing it being equal to the number of Permutations which there are in the fix Letters *abbccc*, it follows, by the Rule before delivered, that the number of thofe Chances will be 60, *viz.* the Fraction $\frac{1 \times 2 \times 3 \times 4 \times 5 \times 6}{1 \times 1 \times 2 \times 1 \times 2 \times 3}$; and confequently that the Probability required will be $\frac{60}{46656}$, and the Odds againft the doing it 46656—60 to 60, or 776 to 1 nearly.

If it were required to find the Probability of throwing Two Aces, Two Two's and Two Three's with 6 Dice, the number of Chances for doing it being $\frac{1 \times 2 \times 3 \times 4 \times 5 \times 6}{1 \times 2 \times 1 \times 2 \times 1 \times 2}$ or 90, and the number of all the Chances upon Six Dice being 46656, it follows, that the Probability required will be expreft by the Fraction $\frac{90}{46656}$.

Again, if it were required to find the Probability of throwing Three Aces and Three Sixes, the number of Chances for doing it being $\frac{1 \times 2 \times 3 \times 4 \times 5 \times 6}{1 \times 2 \times 3 \times 1 \times 2 \times 3}$ or 20, and the number of all the Chances 46656, the Probability required will be expreft by the Fraction $\frac{20}{46656}$.

PROBLEM LIII.

TO *find at* HAZARD *the Chance of the Caster, when the Main being given, he Throws to any given number of* Points.

SOLUTION.

THis being easily reduced to our XLVII*th* Problem, it is thought sufficient to exhibit the Solution of its different Cases in the following Table, which shews the Odds for or against the Caster.

Points Thrown to	MAIN V.	
	exactly	nearly
i	Against the Caster 538 to 407 or	37 to 28.
ii	For the Caster — 989 to 901 or	45 to 41.
iii	For the Caster — 2293 to 1487 or	37 to 24.
iv	For the Caster — 2293 to 1487 or	37 to 24.
v	Against the Caster 2117 to 1663 or	14 to 11.
vi.	Against the Caster 2467 to 1313 or	62 to 33.
	MAIN VI.	
i	Against the Caster 2879 to 1873 or	83 to 54.
ii	Against the Caster 2483 to 2269 or	58 to 53.
iii	For the Caster — 2621 to 2131 or	16 to 13.
iv	For the Caster — 2621 to 2131 or	16 to 13.
v	Against the Caster 2483 to 2269 or	58 to 53.
vi.	Against the Caster 2483 to 2269 or	58 to 53.
	MAIN VII.	
i	Against the Caster 629 to 361 or	7 to 4.
ii	Against the Caster 277 to 218 or	14 to 11.
iii	For the Caster — 251 to 244 or	36 to 35.
iv	For the Caster — 251 to 244 or	36 to 35.
v	For the Caster — 601 to 389 or	20 to 13.
vi.	For the Caster — 263 to 232 or	17 to 15.

MAIN

Points Thrown to	
	MAIN VIII.
i	Againſt the Caſter 3275 to 1477 or 51 to 23.
ii	Againſt the Caſter 2483 to 2269 or 58 to 53.
iii	For the Caſter — 2621 to 2131 or 16 to 13.
iv	For the Caſter — 2621 to 2131 or 16 to 13.
v	For the Caſter — 2483 to 2269 or 58 to 53.
vi.	For the Caſter — 2665 to 2087 or 83 to 65.
	MAIN IX.
i	Againſt the Caſter 2467 to 1313 or 62 to 33.
ii	Againſt the Caſter 2117 to 1663 or 14 to 11.
iii	For the Caſter — 2293 to 1487 or 37 to 24.
iv	For the Caſter — 2293 to 1487 or 37 to 24.
v	For the Caſter — 989 to 901 or 45 to 41.
vi.	Againſt the Caſter 538 to 407 or 37 to 28.

F I N I S.

CPSIA information can be obtained
at www.ICGtesting.com
Printed in the USA
BVHW090429021118
531872BV00007B/290/P

9 781298 559920